Innovation and the Creative Process

NEW HORIZONS IN THE ECONOMICS OF INNOVATION

Founding Editor: Christopher Freeman, *Emeritus Professor of Science Policy, SPRU – Science and Technology Policy Research, University of Sussex, UK*

Technical innovation is vital to the competitive performance of firms and of nations and for the sustained growth of the world economy. The economics of innovation is an area that has expanded dramatically in recent years and this major series, edited by one of the most distinguished scholars in the field, contributes to the debate and advances in research in this most important area.

The main emphasis is on the development and application of new ideas. The series provides a forum for original research in technology, innovation systems and management, industrial organization, technological collaboration, knowledge and innovation, research and development, evolutionary theory and industrial strategy. International in its approach, the series includes some of the best theoretical and empirical work from both well-established researchers and the new generation of scholars.

Titles in the series include:

Technological Systems and Intersectoral Innovation Flows
Riccardo Leoncini and Sandro Montresor

Inside the Virtual Product
How Organisations Create Knowledge Through Software
Luciana D'Adderio

Embracing the Knowledge Economy
The Dynamic Transformation of the Finnish Innovation System
Edited by Gerd Schienstock

The Dynamics of Innovation in Eastern Europe
Lessons from Estonia
Per Högselius

Technology and the Decline in Demand for Unskilled Labour
A Theoretical Analysis of the US and European Labour Markets
Mark Sanders

Innovation and Institutions
A Multidisciplinary Review of the Study of Innovation Systems
Edited by Steven Casper and Frans van Waarden

Innovation Strategies in Interdependent States
Essays on Smaller Nations, Regions and Cities in a Globalized World
John de la Mothe

Internationalizing the Internet
The Co-evolution of Influence and Technology
Byung-Keun Kim

Asia's Innovation Systems in Transition
Edited by Bengt-Åke Lundvall, Patarapong Intarakumnerd and Jan Vang-Lauridsen

National Innovation, Indicators and Policy
Edited by Louise Earl and Fred Gault

Innovation and the Creative Process
Towards Innovation with Care
Edited by Lars Fuglsang

Innovation and the Creative Process

Towards Innovation with Care

Edited by

Lars Fuglsang

Department of Communication, Business and Information Technologies (CBIT), Roskilde University, Denmark

NEW HORIZONS IN THE ECONOMICS OF INNOVATION

Edward Elgar

Cheltenham, UK • Northampton, MA, USA

Published by
Edward Elgar Publishing Limited
Glensanda House
Montpellier Parade
Cheltenham
Glos GL50 1UA
UK

Edward Elgar Publishing, Inc.
William Pratt House
9 Dewey Court
Northampton
Massachusetts 01060
USA

A catalogue record for this book
is available from the British Library

Library of Congress Cataloging in Publication Data

Innovation and the creative process : towards innovation with care / edited
by Lars Fuglsang.
    p. cm.— (New horizons in economics of innovation series)
    Includes bibliographical references and index.
    1. Technological innovations—Management. 2. Creative ability in
business. I. Fuglsang, Lars.
    HD45.I53719   2008
    658.3'14—dc22
                                                                2007029864

ISBN 978 1 84720 387 8

Printed and bound in Great Britain by MPG Books Ltd, Bodmin, Cornwall

Contents

List of figures   vii
List of tables   viii
List of boxes   ix
List of contributors   x

Foreword   xiv
Jon Sundbo

INTRODUCTION

1.   Innovation with care: what it means   3
     Lars Fuglsang

PART 1:   INVOLVEMENT

2.   Innovation and involvement in services   25
     Jon Sundbo
3.   Customer Relationship Management (CRM) as innovation:
     taking care of the right customers   48
     Jan Mattsson
4.   Innovation with care in health care: translation as an
     alternative metaphor of innovation and change   57
     John Damm Scheuer

PART 2:   IMPORTANCE

5.   The public library between social engineering and
     innovation with care   87
     Lars Fuglsang
6.   Getting waste to become taste: from the planning of
     innovation to innovation planning   112
     Gestur Hovgaard
7.   Public innovation with care: a quantitative approach   131
     Lars Fuglsang, Jeppe Højland and John Storm Pedersen

8. Meta-innovations on strategic arenas: innovative
    management in public organizations   142
    Jørn Kjølseth Møller

PART 3:   POSITIONING

9. The interaction between public science and industry, and the
    role of the Øresund Science Region's platform organization   169
    Povl A. Hansen and Göran Serin
10. The role of a network organization and Internet-based
    technologies in clusters: the case of Medicon Valley   193
    Ada Scupola and Charles Steinfield
11. The "Mad Max Puzzle": positioning and the lone inventor   212
    Jerome Davis and Lee N. Davis

PART 4:   SENSEMAKING

12. Sense caring in innovation   237
    Peter Hagedorn-Rasmussen
13. Making innovation durable   254
    Connie Svabo
14. Intrapreneurship: differences in innovations is a matter of
    perspective and understanding   275
    Hanne Westh Nicolajsen
15. Mindful innovation   295
    Poul Bitsch Olsen

Index   311

Figures

8.1.	System of innovation on the domain for educational services in Denmark (kindergartens and services for care of children)	154
8.2.	Diversity and capability to innovate	159
8.3.	Organizational cultures	161
9.1.	The Øresund Science Region	180
11.1.	Initial positioning: Mad Max and Big Widget Inc.	215
11.2.	Mad Max and Big Widget Inc. The impact of expectations	217
11.3.	Time line one: two-step sequencing	219
11.4.	Time line two: multiple step sequencing	219
13.1.	Communication (a)	270
13.2.	Communication (b)	270
14.1.	The innovation process of ProjectWeb	281
14.2.	Screen dump from ProjectWeb (IT department)	282
14.3.	Innovation-in-use?	283

Tables

4.1.	Quantitative and qualitative approach to quality development	71
7.1.	Innovation in public institutions	134
7.2.	The proxy of innovation with care	135
7.3.	Innovation and fulfillment of external demands	136
7.4.	Innovation and strategic involvement	137
8.1.	Institutional pressures on educational organizations	145
8.2.	Examples of strategic arenas in educational organizations (kindergartens)	149
8.3.	Characteristic of networks as a potential and as action (intention)	162
10.1.	Characteristics of Medicon Valley (adapted from the MVA home page, www.mva.org)	202
11.1.	Chronology: *Kearns v. Ford re*: the intermittent windshield wiper	226
14.1.	ProjectWeb content and sense-making conditions across projects	289

Boxes

8.1.	Types of innovations and innovation strategies in the public sector	151
8.2.	Five roles in the management of innovation	158
12.1.	The products offered – and a taste of the unfolding innovation	244
12.2.	The e-realtors' four IT platforms	246
15.1.	Definition of mindfulness	300

Contributors

Jerome Davis is currently Canadian Research Chair (Oil and Natural Gas Policy) at Dalhousie University, Halifax, Nova Scotia. He has published widely in the fields of oil and natural gas policy, and in diverse fields such as equity markets as institutions, the institutional consequences of incomplete contracts, project management and analysis, public sector restructuring, and the role of prizes as incentives to innovation.

Lee N. Davis is Associate Professor at the Department of Industrial Economics and Strategy, and Research Associate at the Centre on Law, Economics and Financial Institutions, both at the Copenhagen Business School. She has conducted research on economic incentives to research and development over the past two decades, with a special focus on the role of intellectual property rights, and published widely in the field. Other research interests include firm appropriability choices, innovation strategy, and academic patenting in the life sciences.

Lars Fuglsang (PhD) is Associate Professor in Social Sciences at the Department of Communication, Business and Information Technologies (CBIT) at Roskilde University. He has written books and articles in the field of innovation studies, public innovation, service development, and science and technology studies. His research explores how organizational frameworks are created to deal with the impact of innovation and technology on business and society.

Peter Hagedorn-Rasmussen (PhD) is Associate Professor of Social Sciences at the Department of Communication, Business and Information Technologies (CBIT) at Roskilde University. He has written books and articles in the field of organizational change, management and work life studies. His main research interest is the study of organizations as living compromises, with particular focus on the relationship between management and work life.

Povl A. Hansen is Dr fil. in Economy History from Lund University, Sweden, and Associate Professor in Economic Geography at the Department of Communication, Business and Information Technologies (CBIT) at Roskilde

University. He has published in the fields of technology, innovation and regional development. His research is in the field of industrial analysis especially focusing on the relationships among innovation processes, industrial structures and transfer of knowledge. He has published many books, reports and articles on technology, business conditions and regional development.

Jeppe Højland is a doctoral student of Social Sciences at the Department of Communication, Business and Information Technologies (CBIT) at Roskilde University. His PhD dissertation is about new kinds of reward practices in knowledge-intensive firms. He has written an article about the impact on leaders and employees of the Danish public sector reform. His main research interest is innovation within the area of human resource management.

Gestur Hovgaard (PhD) is Assistant Professor of Social Sciences at Roskilde University. He has written in the field of innovation and social innovation, and his main fields of interest are within local and regional development. Food chains and food biotechnology are also examined in his research.

Jan Mattsson is Professor in business administration at the Department of Communication, Business and Information Technologies (CBIT) at Roskilde University. He has held several professorships and visiting professorships in New Zealand, Australia and Scandinavia. He has authored several books and more than 50 peer-reviewed international publications in journals such as *The International Journal of Research in Marketing* and *Journal of Economic Psychology*. He serves on many editorial boards of international journals in marketing and services. He takes part in several international research projects focusing on customer-firm interactions.

Jørn Kjølseth Møller (Master of Political Science) is a doctoral student of Social Sciences at the Department of Communication, Business and Information Technologies (CBIT) at Roskilde University. He has written books and articles in the field of strategic management and change, service development and psychology in organizations. His main research interest is in strategic management and change of educational organizations in the public sector.

Hanne Westh Nicolajsen (PhD) is Assistant Professor at the Center for Information and Communication Technologies (CICT) at the Technical University of Denmark. She has published in the field of the use of information and communication technologies in organizations. Her research examines how information and communication technologies are used in organizations and are shaped by organizational and entrepreneurial factors.

Poul Bitsch Olsen (PhD) is Associate Professor of Organization Theory at the Department of Communication, Business and Information Technologies (CBIT) at Roskilde University. He has written books and articles on collective knowing and organizing. In particular, his research examines leadership and project work. His research also explores how collective practicing is the basis for business innovation, team-sports, value production and academic competence.

John Storm Pedersen (PhD) is Associate Professor of Social Sciences at the Department of Society and Globalisation at Roskilde University. He has written books and articles in the field of public administration and management, structural reforms in the public sector and public innovation. His main research interest at present is the impact of the structural reforms in the public sector in Denmark and how public institutions deliver services to the citizens. Pedersen is former CEO of the Mayor's Office in the municipality of Aalborg.

John Damm Scheuer (PhD) is Assistant Professor at the Department of Communication, Business and Information Technologies (CBIT) at Roskilde University. His main research interest and focus is the study of the encounter of innovative ideas and local practice in private as well as public organizations. The encounter is studied as implementation, diffusion or translation processes. But also new and innovative ways of theorizing about "the encounter" and local organizing processes are explored.

Ada Scupola (PhD) is Associate Professor at the Department of Communication, Business and Information Technologies (CBIT), Roskilde University. She has published articles and books in the area of information technology innovation, especially in the field of adoption, diffusion and use of information technologies such as e-commerce and e-services in SMEs and industrial clusters. Her research focuses primarily on how organizational and industrial factors shape the development, adoption, implementation, use, and effects of such technologies.

Göran Serin (Dr) earned his degree in Economic History and is Associate Professor in Business Administration at the Department of Communication, Business and Information Technologies (CBIT), Roskilde University. He has extensive research experience within the fields of technology, innovation and regional development. He has a particular interest in industrial analysis and industrial restructuring and regional development, on which he has published many articles and books. In recent years, his research has especially focused on analysing regional integration in cross-border regions.

Charles Steinfield (PhD) is Professor and Chair of the Department of Telecommunication, Information Studies and Media at Michigan State University. He has published books and articles in the area of organizations and use of information and communication technologies. His research examines how individual and organizational factors shape the development, adoption, use and effects of such technologies.

Jon Sundbo is Professor in Business Administration at the Department of Communication, Business and Information Technologies (CBIT), Roskilde University. He is director of the Center for Service Studies and coordinator of the Department's research area in innovation and change processes in services and manufacturing. He has published extensively in the fields of innovation, service management and the development of the service sector, tourism and organization. He has published articles in several journals on innovation, entrepreneurship, service and management and has authored several books, among these *The Theory of Innovation* and *The Strategic Management of Innovation*.

Connie Svabo (Master in Business Administration) is PhD Fellow at the Department of Communication, Business and Information Technologies (CBIT) at Roskilde University. She has written articles and edited books about practice-based learning. She has several years of professional experience in consultancy work and commercial writing. Her main research interests are organization, materiality and esthetic forms of knowing.

Foreword

This book presents new thoughts and research on innovation. Innovation with care emphasizes both the care for people and the care for doing innovation in a proper way. Most innovative attempts fail and create economic loss and individual disappointment.

Another book in the overwhelming stream of books on innovation? Can it contribute with new knowledge? We believe it can by taking a primarily sociological approach to innovation. Not that the economic aspects are forgotten, but the sociological aspects are emphasized complimentary to the economic ones. This is not very common in innovation literature. The individual – the entrepreneur – has been emphasized, but rarely the social processes with different actors and roles and innovation as an interactive process.

In the increasing contemporary theoretical and practical interest for innovation, the social aspects of the innovation process has often been forgotten. Emphasis has primarily been on economic processes and policy. However, innovation is a process that is carried out by people in interaction with people. It may be that the result of the process is part of the market economy, but the process itself is a social process where the economic results are not at all sure. Recently the social processes have come more into focus. Innovation projects, creativity and user-involvement have become objects in the front research. This book is one contribution to this movement.

The book is a presentation of more than 15 years of research in the Innovation Research Group at Roskilde University in Denmark. In this group we have had a preference for the out-of-mainstream approaches to innovation: Innovation in services and the experience economy, innovation as non-sophisticated, quick practical ideas, continuous incremental innovation, user/customers' and employees' role in the innovation process, innovation as an organizational sensemaking process and so on. This has been amusing and informative for us and we believe it can provide new knowledge for researchers, students and others interested in innovation as both an economic and a social phenomenon.

We think that the future for the phenomenon of innovation – and thus for innovation research and practical innovation work in firms and societies – is to return to the original point of departure: A general change of behavior and economic structures – social and economic change. We believe that innovation in the future will be a much more comprehensive phenomenon than just R&D, entrepreneurship as establishment of new high-tech firms or narrow industrial policy. Social entrepreneurship as solving social problems, innovation as a value creating organizational development factor and as a collective social activity in- or outside the formal economy will probably be future highlights within innovation research. This will develop innovation theory and make it more exciting, but also more diffuse since it will concern social change in general. The latter will challenge the theory development, but there is no way around this if we want to explain economic development, which in the future will concern phenomena such as lifestyle, experience, corporate identity, solution of social problems and so on.

Jon Sundbo
Professor of Innovation and Business Administration
Co-ordinator of the Innovation Research Group
Roskilde University, Denmark

Introduction

1. Innovation with care: what it means

Lars Fuglsang

The purpose of this book is to find new ways to understand and analyse the phenomenon of innovation within the frameworks of "strategic reflexivity" (Sundbo and Fuglsang, 2002; Fuglsang and Sundbo, 2005) as well as "open innovation" (Chesbrough, 2003). The book presents new insights into mechanisms that are important for benefiting from innovation across sectors, organizations and people. It deals with tensions and paradoxes in innovative activities between, for example, variety and selection, creativity and innovation, or between business innovation and social innovation – rather than seeing innovation from one particular point of view. "Innovation with care" means that innovation is seen as something that takes place among many actors having different perspectives, ideas and cultures that have to be carefully woven together in order to achieve the benefits of innovation.

We understand innovation as an interactive process that involves many people and often also changing people across sectors. Innovation is therefore a common activity, which is not restricted to special groups of persons such as "the creative class" or "symbol analysts" or people working in R&D labs. Innovation is a process that increasingly engages ideas and opinions from many different people. These are opinions and ideas that have to be expressed but also selected and aggregated. Innovation requires diversity and collectivity at the same time, and the balance between the two is a crucial aspect of innovative activities today. This balance is affected both by the market and by other social and organizational forces.

We examine how social and organizational forces are important to that balance and these tensions. In doing so, the book tries to distinguish between different organizational and societal mechanisms of diversity and collectivity, especially four. These are: involvement, importance, positioning and sensemaking. These all operate on different levels (micro and macro) that can be understood and analysed from a sociological and economic perspective (see later).

Involvement is a mechanism of diversity mostly at the organizational level, where people can deliberately involve others in an exchange of experiences and considerations of workable ideas. Involvement presumes that those who are involved are relatively independent and can speak freely about their opinions.

Importance is a mechanism of collectivity. It presumes that certain initiatives are seen as better than others and therefore are exposed and diffused more widely, so that people can adopt them or learn from them. The notion of importance also implies that we are here not only speaking of the market mechanism, but also of mechanisms that involve an element of voice (not just choice). Hence, importance is a social mechanism of aggregation, selection and diffusion.

Positioning is a mechanism of diversity at the societal level. It requires that people have the freedom to pursue their economic and social interests and can express their opinions about ideas they perceive to be socially effective or desirable. According to this mechanism, people can position themselves as actors in innovation and economic development.

Finally, sensemaking is a mechanism of collectivity at the micro-level, where people try to make sense of their experiences and thereby to discover, not so much what is perceived as important and wise in the larger context, but what is appropriate and meaningful in a specific context.

The different chapters of the book deal in different ways with these crosscutting aspects of innovation. Some have a stronger focus on the macro-level and some on the micro-level. Some tend to stress diversity and others collectivity. Therefore, the chapters have been grouped in different sections entitled involvement, importance, positioning and sensemaking. Nevertheless, what binds together all of the chapters is the attempt to see innovation from a broader perspective and at the systemic level where the tensions between actors, and between diversity and collectivity, can be studied.

BACKGROUND

Innovation with care is generally an approach to innovation that starts from the following premises:

1. that innovation and the way in which innovation takes place is important to economic growth and social development;
2. that the concept of innovation has to be better understood in terms of how it can be applied in practice, especially through case-studies; and
3. that innovation in practice requires a reflexive approach that takes into account both economic and social elements, as well as tensions across sectors, organizations and people.

Innovation with care grows out of research undertaken at Roskilde University over recent decades. This research has focused upon how innovation is changing from a technological and industrial mode to a reflexive mode involving many types of institutions, sectors, companies and social groups (Fuglsang and Sundbo, 2005; Sundbo and Fuglsang, 2001). This calls for a new conceptualization of innovation as well as social development, which can take into account the heterogeneity of relationships that evolve around innovative activities.

Today, the innovative resources are much more widely distributed throughout society than just a few decades ago (Chesbrough, 2003). Innovation is no longer based in companies' R&D departments, or in the state's large-scale projects. Innovation can be understood as an interactive process that involves many and changing actors over time, and which serves multiple concerns and conglomerates of different users. This also makes it more challenging for people to integrate different ideas and opinions about innovation, to balance goals and means, and to create frameworks of mutual communication, collaboration and understanding. It becomes critical to analyse how this heterogeneity among sectors, organizations and people can be managed in different ways, and in different social and strategic arenas.

This historical approach to innovation can also be seen as being opposed to a more homogeneous approach to innovation where innovation is seen as something which is planned and managed in a more straightforward and detailed way. The heterogeneity of innovation today means that innovative activities cannot be easily controlled through detailed planning, but that opportunities for innovation have to be continuously evaluated, interpreted and interfered with. Innovative organizations become interpretative systems (Daft and Weick, 1984; Fuglsang and Sundbo, 2005). They try to create some sense of direction and integrate people into changes by making interpretations about the changing opportunities of the organization, and how skills and opportunity can be adjusted to each other.

The main question, which should be explored, is therefore how a proper environment for innovation can be constructed that takes into account these complex mechanisms of diversity and collectivity. Furthermore, an approach that uses various types of case studies in combination with other research techniques may turn out to be an important way in which such mechanisms can be better understood. It requires a more problem- and action-oriented approach to the study of innovation.

The purpose of the book is to demonstrate that this approach is a fruitful approach to innovation. Along these lines, innovation is, in our view, more about interpretation than planning, more about heterogeneity and tension than homogeneity and control, more about opinion than choice, and more about wisdom than science. The book will illustrate the value of innovation with care through a number of interesting cases.

CREATIVITY AND INNOVATION

In most definitions of innovation, innovation consists of two aspects: creativity and innovation (Amabile *et al.*, 1996), invention and diffusion (Rogers, 1995), exploration and exploitation (March, 1991), or variation and selection (Nelson and Winter, 1977). This means that innovation is seen as consisting of two integrated processes. New appropriate ideas or inventions have to be explored in a creative way. In addition to this, there is another parallel process going on where these ideas are aggregated, selected, diffused, implemented and exploited. The interaction between these two sides of innovation is critical.

For example, Teresa Amabile, who has studied creativity, defines innovation in the following way. "We define innovation as the successful implementation of creative ideas within an organization. In this view, creativity by individuals and teams is a starting point for innovation; the first is necessary but not sufficient condition for the second" (Amabile *et al.*, 1996: 1154–5).

Mulgan and Albery who have studied public and social innovation have defined innovation in a similar vein: "We define innovation as 'new ideas that work'. To be more precise: Successful innovation is the creation and implementation of new processes, products, services and methods of delivery which result in significant improvements in outcomes efficiency, effectiveness or quality" (Mulgan and Albury, 2003: 3).

And Sundbo (1998: 12) in a book about service innovation gives the following definition: "I will use 'innovation' to describe the effort to develop an element that has already been invented, so that it has a practical-commercial use, and to gain the acceptance of this element."

Most of these definitions of innovation have their origins in Schumpeter's original definition of innovation in his *Theory of Economic Development*. In that book, Schumpeter is, among other things, conceiving the function of the entrepreneur for economic development as someone who goes against the mainstream. In his definition of innovation, Schumpeter stresses both the ability of the entrepreneur to create entirely "new combinations" and to open up new markets and teach the consumers to use innovations (Schumpeter, 1934, 1969: 65–6).

This double sidedness of innovation is at the core of this book as it is also the core of many descriptions of creativity. For example, in his book about creativity, Csikszentmihalyi makes a distinction between the domain of creativity and the field of creativity (Csikszentmihalyi, 1996). The domain is the specific area in which someone is creative, for example in science, art or politics. The field is the wider environment in which creativity is recognized and stimulated. It is often the field of innovation, rather than the domain, which explains, according to Csikszentmihalyi, why some organizations are more creative than others, for example why Florence was a particular creative city in the Florentine renaissance.

On the other hand, in this approach, we should not ignore the critical factors that are important to creativity or the social psychology of creativity. For example, in her work on the social psychology of creativity, Teresa M. Amabile has examined the proposition that intrinsic motivation is crucial to creativity (Amabile, 1996). Creativity is, as Schumpeter explained, an act that is often motivated by itself rather than by external requirements or extrinsic motivations. Creativity also requires that people are situated in a domain where they possess the necessary domain-relevant skills – rather than being put into a domain where they possess no such skills.

To treat creativity with care means that people are not moved to a domain where they possess no domain relevant skills, and that intrinsic motivation is not entirely replaced by extrinsic motivation. This may increase the chances that creative results are appropriate and not bizarre or eccentric. In this way creativeness means the ability to create new innovative results that are meaningful and appropriate. Understood in this way, the right microenvironment for creativity increases the chances of creative ideas that are also appropriate to innovation.

BUSINESS INNOVATION AND SOCIAL INNOVATION

Innovation with care is also an approach to innovation which focuses upon the interdependencies that exist within social and economic development between business innovations on the one hand, and "social innovations" on the other – in addition to the interaction of creativity and innovation. In the perspective of social innovation, the relation between creativity and innovation, variation and selection, gains another meaning, as we shall briefly explore in the following.

Social innovations are innovations based in social goals and social processes. Social innovation (Young Foundation, 2006) and business innovation, however, often overlap and intertwine. A mobile telephone is a social innovation and a business innovation at the same time: it serves a social goal for those who use it, and it generates economic value for those who produce it and for society. A new model of a mobile telephone is perhaps not necessarily a social innovation if it does not serve a particular (new) social goal. To understand and analyse the social goal and value of an innovation can be an important driver of business innovation.

Another aspect of social innovation besides the social goal and social value is the social process and diffusion of innovation. All processes of innovation, business innovations, social innovations or innovation with care, could, as indicated, be said to consist of two related processes: exploration and exploitation (March, 1991), variation and selection (Nelson and Winter, 1977), or invention and diffusion (Rogers, 1995). Ideas are developed, and then some of these ideas are selected and scaled up as more important or wiser than others. The selection mechanism can be "the wisdom of the crowds" (Surowiecki, 2004), or it may be opinion-makers or society's elite that attempt to "pick the winner".

Seen from the perspective of business innovation, the market mechanism is crucial to this tension and the transition between invention and diffusion. The "wisdom of the crowd" is here expressed by consumer choice implying that certain ideas are picked by the crowd as being more effective or in some sense better than others. Seen from the perspective of a social innovation and innovation with care, the market is not the only mechanisms of diffusion and scaling up. Here, other social mechanisms of aggregation, selection and diffusion are important too (see Rogers, 1995), and these are sometimes more difficult to come to grips with from an analytical point of view.

In some cases, as in public innovation, the market mechanisms may be entirely missing, and therefore the social mechanisms of aggregation, selection and diffusion of innovation present an even more challenging task when analysing it at the theoretical level and constructing it at the practical level. For example, the market mechanism cannot always be applied to the selection of adequate learning or teaching tools in schools. Here, inputs from many professionals and experts and even systemic reviews and demonstrations of these inputs may be needed as a basis for selection. In other cases, the social selection and diffusion of innovation is intertwined with some kind of market mechanism, where the behavior of the user and user choice in combination with user voice may be relevant. This is typical for public television for example.

Innovation with care is therefore an approach to innovation that tries to pay more attention to social mechanisms of diffusion and scaling up than is usual, and to analyse the complex interactions that take place between market mechanisms and social mechanisms, understood as a mechanism of selection and diffusion.

In some cases, business innovations and social innovations, as mentioned, strongly overlap. In other cases social innovations are separate domains but still crucial to business innovation. The university is a social innovation in its own domain, which still plays a crucial role for economic and social development more broadly. Health is dependent on development of commercial drugs as well as a range of social innovations including the health system itself. The sewerage system is a social innovation that is crucial for health conditions. Wikipedia is a social innovation that may end up as a commercial innovation, but initially it started out as a social idea. Social innovations, such as a library or a laboratory, provide social structures that inform the chances of discovery, creativity and innovation more broadly in society. In some cases, social innovations are therefore domain specific, in other cases they provide a different perspective on business innovation, as in the case of the mobile phone or the use of SMSs.

The practical relevance of this approach is that there is a difficult balance to maintain within social and economic development between market mechanisms on the one hand, and social mechanisms of innovations on the other. Furthermore, it becomes important to investigate how social mechanisms can be constructed. More energy and resources could be devoted to the study of the social sides of innovations because they are crucial to innovations both to the chances of creative discoveries and to the robust and meaningful application and implementation of these discoveries in business and society. As such, social mechanisms of innovation also inspire economic development more broadly and can be pertinent to the implementation and the commercial exploitations of innovations. Innovation with care is in this way really a broad concept for the productive tensions and balance between the commercial and the social sides of innovation.

Innovation with care is also an approach to innovation which is different from social engineering. Innovation with care represents, as we define it, an incremental way to improve society and its institutions, and is based in the social values and independent opinions of many citizens in the context of, for example, a nation, that have to give their support to it. It corresponds more or less to Karl Popper's idea of "piecemeal" social change as opposed to utopian social engineering, hence it is a critical-reflective and incremental approach to innovation (see Popper, 1962).

Innovation with care is a complex and a time-consuming process where much can go wrong, because of the many unintended consequences and risks associated with innovation. Therefore, we need a careful approach to innovation where many perspectives and ideas are woven together in a careful way.

Another important issue here is that innovation can be motivated in many ways as Schumpeter already pointed out in his *Theory of Economic Development*. Profit is one motive, but the creation of a small kingdom or the wish to solve concrete problems could be other reasons for people to engage in innovation. Many important innovations cannot, initially, be, in a meaningful way, understood directly as commercial innovations, driven by profits, since their "business model" sometimes is very unclear. Google and Wikipedia are two prominent examples. Other examples are those development projects that take place in the name of cultural policies. For example, when politicians, volunteers, business leaders and social entrepreneurs go together to develop their municipality, the direct commercial spin-offs are usually very difficult to calculate and understand. This kind of innovation is better described as a social innovation of its own purpose and in its own right that may be complementary to business innovations in the municipality, for example the well functioning of the labour market and the chances of creativity and meaningful innovation.

Social innovations such as schools, universities, libraries, and so on can be driven by public or private organizations. Voluntary organizations or enthusiasts as well as business organizations can also drive social innovations, and sometimes we tend to forget the significance of these initiatives. For example, in the case of the so-called information society, the field of social innovations is often much more important to discovering needs and socially effective solutions to them than is normally understood. Social or public entrepreneurs and social institutions can play a crucial role for developing network technologies to people that eventually become real resources to them. These kinds of innovations can inspire business innovations and can lead to the development of highly meaningful commercial innovations. But the field of social innovations is very poorly understood today in comparison with many other high-profile commercial innovations. There is a hidden economy of social and volunteer innovation at risk here, which needs to be explored in order to identify good examples of socially effective ideas that others might be able to learn from.

A main difficulty within the field of social innovation and innovation with care is, however, that the impact of this approach is difficult to define and measure. For example, what is the impact of the university? What is the impact of the library and the school? It can be explained in broad terms, but it is much more difficult to measure and analyse than commercial innovations. Indeed, in some cases, certain success criteria can be created, such as the impact of the health system or a particular medical treatment on life expectancy. Still, even in this case, the complex interaction of treatment, sanitary conditions and nursing are very difficult to explain. And in many other cases, the measurement of impact is a very difficult issue. Furthermore, the impact of social innovation and innovation with care on business innovation is an extremely complicated issue in itself. For example, what is the impact of the public library on business innovation? Does it in one way or another improve the quality of the work force?

But these difficulties of measuring the impact of social innovation should not mean that we neglect to throw more light on these issues and their critical impact on society.

SCHUMPETER I, II AND III

Our approach to innovation is partly inspired by the tensions in Joseph Schumpeter's work between his early work, so-called Schumpeter I, and the late work, so-called Schumpeter II (see Phillips, 1971 for a discussion). One might say that our approach is an attempt to formulate a Schumpeter III approach that reconsiders the wider context of innovation.

In the early work, Schumpeter stressed the role of the entrepreneur (Schumpeter, 1934, 1969). The entrepreneur is described as a special type of person with a special motivation, which is not necessarily driven by profits. The entrepreneur is a dynamic person outside the mainstream, and has a function for changing economic structures. In his later work, by contrast, Schumpeter thought that the social function of the entrepreneur had disappeared (Schumpeter, 1947). He thought that innovation had become a routine-activity in the labs of the modern business corporation. Schumpeter claimed that the entrepreneur was no longer a very relevant type since innoation had become more integrated into society.

The particular distinction between the entrepreneur and routine-based innovation in Schumpeter's work is important here. But what also matters is that Schumpeter's work tells us that different forms or modes of innovation exist, such as the entrepreneurial form in the late nineteenth and early twentieth centuries and the routine-based form in the mid-twentieth century. Furthermore, there may be different motives and rationales behind the various approaches to innovation as well.

What we are missing in Schumpeter's work is, perhaps, a recognition that these two forms of innovation may exist at the same time and in some ways are interdependent. Schumpeter, by contrast, seems to think that one historical period (that of heroic entrepreneurs) is being entirely replaced by another (that of the large corporations). Clearly, from what we know today, this is wrong. There is still a function for entrepreneurs, and today new forms of entrepreneurship are being promoted, such as social entrepreneurship and public entrepreneurship. They have, however, to be understood at the systemic level rather than at the level of the individual.

The distinction between Schumpeter II and Schumpeter III is of the same character as the distinction between Schumpeter I and Schumpeter II. These three approaches represent different frameworks of innovations, where innovations proceed in different ways, are motivated by different factors, aggregated, selected and diffused in different ways, and often crystallize into different kinds of institutions with different underlying rationales.

What we argue, along with Chesbrough (2003), is that innovation today is no longer only based in the routines of R&D inside large corporations. But we also argue that this leads to the formation of new mechanisms of creativity and diffusion of innovation that can be understood at the systemic level. In the field of science studies, Nowotny and others have been discussing so-called mode II science, where external requirements come to play a growing role (Nowotny *et al.*, 2001). They use the metaphor of "agora" to describe how a public arena of science is formed where the quality and the relevance of science is being discussed by scientists as well as by citizens and politicians. In the same way we argue that a new mode of innovation is emerging, the Schumpeter III approach, where new mechanisms of creativity and diffusion of innovation are becoming important, and where market mechanisms and social mechanisms are blended in new ways. For example, we argue that new strategic arenas of innovation are being formed to which many types of actors are linked, such as universities, companies, government institutions and user groups. Furthermore, employees and consumers are interlinked in new ways at the systemic rather than only the individual level.

One important expression of Schumpeter III that we want to stress in this book is innovation in services. Service providers are very dependent on their front personnel and their ideas, because the services are often co-produced by the consumer and co-consumed by the provider, and the concrete work that goes into this is often difficult to describe in very precise terms. Some innovations in services are indeed business innovations, but many innovations may be better explained as social innovations in the sense that they are more related to the problem-oriented work of the front personnel than directly to a commercial strategy. It is based in the interpretation and understanding of a situated problem. This problem-orientation can also often be interlinked with the life-story, intrinsic motivation and professional pride of the service-worker.

Of course, the clever service provider will try to involve the service worker as much as possible in the business development, because he wants "ideas that work" (Mulgan and Young) drawn from the experience of the service worker. This becomes even more crucial today under the new 'Schumpeter III paradigm.' The employee and the consumer therefore become more and more involved at the systemic level in innovation activities, rather than they are interacting and developing services only in individual face-to-face-relationships.

Both in the private and in the public sector, the "journey to the interface" becomes an important theme (Parker and Heapy, 2006), that is the journey to the interface between the front personnel and the consumer. Experiences in the front have to be explored and collected in a more systematic way. In this journey, the interpretation and understanding of the needs of the consumer and the social goals involved in consumer behavior in a broader sense becomes crucial. In the "journey to the interface" the knowledge and experience of the front-personnel is therefore also critical. At the same time, the front-personnel are seen as people who should not act on their own, but on behalf of the system (see Parker and Heapy, 2006). In this way, the issue becomes one of how people can learn from each other and how the good ideas can to be scaled up and diffused across and between people in organizations. This systemic orientation towards external ideas and their combination with internal experiences is an important aspect of the Schumpeter III model.

The diffusion of the Internet and WEB2 and the many new services that are offered on the Internet provide plenty of examples of innovation with care where the innovative resources are highly distributed and where some innovative ideas are being systematically collected and scaled up, or picked by the crowd or through other mechanisms of selection. For one thing, these services are often developed as social innovations in a community of practice and then eventually they develop into business innovations. Facebook is an example of this (see Ellison *et al.*, 2006). It was initially a network technology inspired by earlier ways to introduce students to each other using photographs in a physical "facebook". Then it was developed by a student at Harvard University into a worldwide e-based service for students. It will probably evolve into a commercial innovation in time to come.

Schumpeter III is a mode of innovation where the interpretation, exploration and exploitation of external ideas as well as their combination with internal experiences and ideas become increasingly important (Chesbrough, 2003), and where the diffusion and scaling up of ideas that are thought to be better take new forms in the market as well as society and in organizations, as we shall explore in this book.

DIVERSITY AND COLLECTIVITY, VARIETY AND SELECTION

Innovation requires diversity and collectivity, variety and selection, creativity and innovation at the same time. The balance between the two sides of the coins is a crucial aspect of innovation – and crucial to this book. This balance we understand as affected both by the market and by other social and organizational forces.

The book tries to distinguish between different analytical perspectives that can be helpful for studying these balances or tensions. As a way to organize the chapters of the book we can discern four broad analytical perspectives. These are as previously mentioned: involvement, importance, positioning and sensemaking. These should be understood as different broad perspectives on diversity and collectivity on different levels. What follows is a brief description of these analytical perspectives that organize the chapters of the book after which the single chapters will be presented.

Involvement

Involvement means that innovators can seek to involve many opinions and ideas during innovation. Often, for innovation to take place, it is important that employees and users are involved in the exploration of inventions and new "ideas that work." How this involvement can take place is a complicated issue and a fruitful ground for new research as well as case-studies.

Involvement is a mechanism of diversity or variety, but it also requires that management carefully selects some of the ideas while others are dismissed. Hence, involvement requires a careful approach to both variation and selection, or, as it is explained in Sundbo's chapter, to reflexivity and strategy-making at the same time.

Most obviously, employees can be involved in innovation activities. But consumers can also sometimes be involved. For one thing, they can be involved through the employees having many years of experience with consumers. The employees' discovery of consumer needs can sometimes be crucial for improving goods and services. This is true especially in services and public services.

In some cases, the exploration of consumer needs may be more difficult than in others. For example, in the public sector, a principle of universalism is often important, and employees for good reasons have to think in terms of rules and public law rather than individual needs. To listen more carefully to individual citizens or to make use of employees' experiences with them may almost constitute a paradigm shift in the public sector.

Furthermore, while in many settings involvement of employees and consumers may work in the individual case, in the changing context of innovation, the involvement of employees and consumers must, as mentioned, increasingly take place in a systemic way. People must learn to act on behalf of the company system rather than on behalf of themselves. This also requires a careful balancing of strategy and reflexivity.

Importance

Importance stresses the principle that some ideas may tend to become more widely diffused than others. Importance therefore refers to a mechanism of collectivity and selection. It presumes that certain initiatives are perceived as better or more appropriate than others and therefore are recognized by more people, and that people are willing to adopt them or learn from them.

The notion of importance also implies that we are not only speaking of the market mechanism, but also of a mechanism that involves an element of voice. The market mechanism cannot always be used to scale up and diffuse important ideas. Thus, in many institutional settings, and in the context of many services, people cannot, in practice, make use of the market mechanism, because they are dependent on the services that are provided where they live and work, such as schools, kindergartens, restaurants, local cultural offers and so on.

To find ways other than the market mechanism to promote experiences, make visible the good ideas and scale up the better initiatives so that others can learn from them is a major challenge for many social and public services.

Positioning

Positioning means that companies and institutions can position themselves as actors in economic change and innovation. They are not role-takers in a passive way, but they can actively position and reposition themselves in relation to each other. Nevertheless, in the context of increasing complexity, it may sometimes be quite challenging to acknowledge and recognize each other's competences and qualities.

Positioning is a mechanism of diversity or variety at the societal level. It requires that people have the autonomy to pursue economic and social interests and can express their opinions about ideas they perceive to be relevant – and can communicate what their own contribution to innovation and development may be.

Positioning can be thought of as something that takes place both among individual persons and among institutions. For example, classical entrepreneurship is an embodiment of positioning, where individual persons position themself in relation to other people. But also institutions, such as universities, schools and libraries, have a need to position themselves in order to demonstrate their value to others. Hence, the university is not the same as an R&D lab in a private firm and the public library is not the same as Google (two examples from this book). They each have to position themselves in order to make clear what their individual gift is. How can they do this?

Sensemaking

Sensemaking means (following Weick, 1995) that there must be room for continuous sensemaking in connection with innovation processes in order for people to discover and exploit new experiences and ideas and make them intelligible among each other. Managers and employees must create a "mindful environment" where they collectively can make sense of people's changing perceptions and ideas of social and economic opportunities.

Sensemaking is a mechanism of collectivity or selection at the micro-level, where people together try to make sense of changes and thereby to discover, not so much what is perceived as important and wise in the larger context, but what is appropriate and meaningful in a specific situated context.

A metaphor for sensemaking, which is used in this book in several chapters, is that of "translation." Translation means that new ideas, goods and services are translated or transformed by people to fit the local context before they can be used. This is critical in the emerging context of open innovation and strategic reflexivity. External ideas have to be adapted to the local context before they can be used, and sometimes, when it is difficult to "translate" them, they must be dropped.

Perhaps this is also an approach to innovation that has a particularly strong hold on Scandinavian societies. Here the adoption of ideas, inventions and technologies from the outside world, and the attempt to translate them into something locally valuable, has been a critical aspect of economic development and the development of the welfare states.

THE CHAPTERS OF THE BOOK

All of the chapters in the book deal with certain tensions or paradoxes in innovation, as described above, which relate to involvement, importance, positioning, or sensemaking. Innovation with care is thus an approach that seeks to analyse and understand how people in empirical cases are dealing carefully with these tensions and paradoxes.

In the first section of the book about "involvement," we investigate tensions between engaging people's opinions and ideas on the one hand, and the overall strategy of a company or an organization on the other.

Jon Sundbo examines tensions in organizations between the involvement of employees in innovation, and care for the overall strategy process of a company or an organization. Sundbo pays attention particularly to innovation in services, and his chapter includes a review of the literature on service innovation with respect to the involvement of employees. Sundbo's chapter also draws on a multiple case approach to service firms. Sundbo concludes that service innovations are based on care for the strategic reflexive processes as well as the actors and roles involved in service production. This care does not only mean the nursing of the people and encouraging intrapreneurship, but also restrictions and a strict decision process concerning new ideas and innovation projects.

Jan Mattsson examines tensions in an organization between different facets of care that are important to working with Customer Relationship Management (CRM). Taking care is the physical handling of the CRM innovation process and running the subsequent CRM system. Caring for customers and employees is a psychological sensitivity to how customers and employees react when faced with change and re-organization. Careful operation means that safeguards are in place and care is taken when designing the system to ascertain the increased value can be offered to customers in exchange for the extra effort of data input and co-ordination.

John Damm Scheuer explores tensions that emerge in an organization when a general idea travels into the organization and is translated into something useful by the people inside that organization. In a case study of innovation in health care (the case of the "clinical pathway"), he argues that the innovation process may be better understood if theorized as a translation rather than an implementation or rationally planned process. Scheuer also argues that the concept of "innovation with care" may be defined as local translators' translation of innovative ideas in a way that tests the pros and cons of an idea in relation to local knowledge and takes appropriate steps to integrate those elements of the idea.

The second section of the book is about "importance." It examines how certain ideas are scaled up and selected and in some sense become more important than others, in a careful balancing of variation and selection, or exploration and exploitation.

Lars Fuglsang explores some of the tensions between variation and selection in the context of public innovation. Through a case study of the Danish public library, he tries to build an analytical framework for analyzing innovation with care in the public sector. He shows how the mode of innovation is partly changing from "institutional innovation" to "open innovation." This leads to a quest for mechanisms of variation and selection, rather than, for example, mechanisms of homogenization among public institutions. Fuglsang argues that new social and strategic arenas are created in the library case, which enables variation and selection, and the diffusion of important new ideas.

Gestur Hovgaard examines tensions that exist between exploration and exploitation, as well as stability and change, in the construction of an innovative new company, which is the Danish food-ingredient company Danmark Protein (DP). Today it is incorporated within the dairy giant Arla Innovation. Hovgaard shows how exploration and exploitation are activities that vary, due to different modes of innovation. This is a similar argument to the one made by Fuglsang in his chapter. Finding a proper balance between exploration and exploitation is a key to the success of a company. This requires an "interpretative tradition," mutual networking and a common understanding. This is consistent with both Sundbo and Fuglsang in this volume.

Lars Fuglsang, Jeppe Højland and John Storm Pedersen applies a quantitative analysis to investigate tensions between variation and selection again in the public sector. They present a survey, which has been sent to leaders in Danish public institutions about innovation activities. The chapter seeks to define a proxy for innovation with care in order to quantify its impact on various output measures. The survey shows that innovation does take place in the public sector, and that variation rather than copying is the rule. This is consistent with Møller's findings (see below). The chapter also documents that innovation with care is an effective way of dealing with external requirements.

Jørn Kjølseth Møller examines tensions between path-dependency and diversity in the public sector with a special view to a "Management Greenhouse" created by employers and employees' organizations in Danish municipalities. He argues that diversity in the public sector is a more common phenomenon than normally understood in, for example, neo-institutional theories. He shows how the potential for innovative management in public organizations is determined by the institutional context, where the public institutions are functioning. Similarly to Hansen and Serin (see below) and Fuglsang (the arena approach) he argues that these issues are incorporated into strategic arenas, where specific types of interests are expressed, specific issues negotiated and specific rules of the game established. The Management Greenhouse is an example of this.

In the third section of the book, three chapters discuss how actors can position themselves in various ways in the broader, macro-economic context of innovation – and what tools are available for that.

Povl A. Hansen and Göran Serin investigate tensions between universities and firms and different notions of public and private research. By way of a case study of Øresund Science Region they argue that a "platform organization" can be seen as a solution to these tensions when universities and firms have different perspectives on the purpose and structure of research. According to Hansen and Serin, a platform organization can promote a caring approach among the different institutions to each other's approaches. Hansen and Serin use the so-called triple helix model of government-university-industry interaction to explain the role of the platform organization as a framework for interaction and positioning.

Ada Scupola and Charles Steinfield explore the tensions that exist between firms' globalization and localization perspectives, and how these differing perspectives can be "taken care of". Through a case study of Medicon Valley, a leading biotechnology cluster in Denmark, they show how Internet-based technologies and a number of additional critical activities can contribute to and support the development of localized economies such as industrial clusters, while also contributing to the globalization of the economy by connecting companies and clusters of companies across different regions of the world.

Jerome Davies and Lee N. Davies examine tensions between an individual inventor (the "Mad Max") and the commercial context in which that inventor has to position himself. The chapter refers to a case study of inventor Robert Kearns and his lawsuit against the American and European automobile industry. Davies and Davies argue that many inventions may not have any commercial potential to begin with, irrespective of what the "mad" inventor may have thought. Although inventors are "ripped off," this may be more a reflection of their lack of positioning skills than of any major wrong-doing on the part of their financial partners. Those inventors who have been "ripped off," as was the case with Kearns, might have avoided this fate by observing the signals of their opposite number more "carefully."

In the final section of the book about "sensemaking," four chapters examine tensions and paradoxes at the micro-level that are critical to benefiting from innovation.

Peter Hagedorn-Rasmussen analyses tensions between strategy and sensemaking by way of a case-study of a e-realtor company under creation. "Care" describes, according to Hagedorn-Rasmussen, an approach that bridge the relationship between the seemingly uneasy pairs of strategy and sensemaking. Care implies a very broad range of meanings including assiduousness, thoughtfulness, sensitivity, consideration but also anxiety, trouble and concern. It might be argued that this lack of conceptual accuracy makes it an odd concept in (micro)sociological studies. On the other hand, the connotations we attribute to the concept of care may be highly accurate and descriptive for the processes of innovation as well as entrepreneurship, where the balance between strategy/strategizing and sensemaking is important.

Connie Svabo explores tensions between innovation and durability in an innovation project in the fashion industry. She argues that an innovation, paradoxically, is an artifact, which is continuously engineered and maintained in a stable form. She presents a case study (or story) of an innovative project, "Sidecar" in a small-scale fashion industry, which was both a success and a failure. In line with the approach presented by Scheuer, Svabo focuses on actor-network theory and translation. The chapter tells a story of the struggles of translating innovative ideas into material forms, and shows that the work of innovation consists of continuous attempts to create material order.

Hanne Westh Nicolajsen examines tension between a new networked communication technology, called ProjectWeb, and three organizational settings in which it is implemented. Her study demonstrates the critical role of the individual entrepreneurs (or "intrapreneur") as an integrating force of technology, innovation and organizational change. According to her study, entrepreneurs are not only needed in the initial phase of idea generation, but also in the phase of implementation. Hanne Westh Nicolajsen argues that it is extremely important to make sure that at least one central person have the right qualifications and interests in order to benefit from technological changes and innovations in an organization.

Finally, Poul Bitsch Olsen, inspired by the approach of Karl Weick, analyses a tension in innovation projects between old and new experiences. His example comes from sports: the continuous innovation that goes between a handball league coach and his team. Olsen shows how the experience of interruptions and change processes must be continuously made intelligible and selected at the micro level. He argues that "mindfulness" or "mindful innovation" is a concept that can be used to understand how creative action and new experiences are carefully selected and used in this way. Mindfulness in innovation means that experience is noticed and made intelligible, and new knowledge is the outcome of this.

REFERENCES

Amabile, T.M. (1996), *Creativity in Context: Update to The Social Psychology of Creativity*, Boulder, CO: Westview Press.

Amabile, T.M., R. Conti, H. Coon, J. Lazenby and M. Herron (1996), "Assessing the work environment for creativity," *Academy of Management Journal*, **39** (5), 1154–84.

Chesbrough, H.W. (2003), *Open Innovation: The New Imperative for Creating and Profiting from Technology*, Boston, MA: Harvard Business School Press.

Csikszentmihalyi, M. (1996), *Creativity: Flow and the Psychology of Discovery and Invention*, New York: HarperCollins.

Daft, R.L. and K.E. Weick (1984), "Toward a model of organizations as interpretation systems," *Academy of Management Review*, **9** (2), 284–95.

Ellison, N., Ch. Steinfield and C. Lampe (2006), "Spatially bounded online social networks and social capital: the role of facebook", paper read at The Annual Conference of the International Communication Association (ICA), 19-23 June, at Dresden, Germany.

Fuglsang, L. and J. Sundbo (2005), "The organizational innovation system: three modes," *Journal of Change Management*, **5** (3), 329–44.

March, J.G. (1991), "Exploration and exploitation in organizational learning," *Organization Science*, **2** (1), 71–87.

Mulgan, G. and D. Albury (2003), *Innovation in the Public Sector*, London: Strategy Unit, Cabinet Office.

Nelson, R.R. and S.G. Winter (1977), "In search of useful theory of innovation," *Research Policy*, **6** (1), 36–76.

Nowotny, H., P. Scott and M. Gibbons (2001), *Re-thinking Science: Knowledge and the Public in an Age of Uncertainty*, Cambridge: Polity Press.

Parker, S. and J. Heapy (2006), *The Journey to the Interface*, London: Demos.

Phillips, A. (1971), *Technology and Market Structure: A Study of the Aircraft Industry*, Lexington, MA: Lexington Books.

Popper, K.R. Sir (1962), *The Open Society and its Enemies*, London: Routledge & Kegan Paul.

Rogers, E. (1995), *Diffusion of Innovations*, New York: Free Press.

Schumpeter, J.A. (1934, 1969), *The Theory of Economic Development*, Oxford: Oxford University Press.

Schumpeter, J.A. (1947), *Capitalism, Socialism, and Democracy*, London: Allen & Unwin.

Sundbo, J. (1998), *The Organisation of Innovation in Services*, Frederiksberg: Roskilde University Press.

Sundbo, J. and L. Fuglsang (eds) (2002), *Innovation as Strategic Reflexivity*, London: Routledge.

Surowiecki, J. (2004), *The Wisdom of Crowds: Why the Many are Smarter than the Few and how Collective Wisdom Shapes Business, Economics, Societies, and Nations*, London: Little Brown.

Weick, K.E. (1995), *Sensemaking in Organizations, Foundations for Organizational Science*, Thousand Oaks, CA: Sage.

Young Foundation (2006), *Social Silicon Valleys: A Manifesto for Social Innovation, What it is, Why it Matters and How it can be Accelerated*, London: Young Foundation.

PART 1

Involvement

2. Innovation and involvement in services

Jon Sundbo

INTRODUCTION

This chapter will discuss innovation in services and employees' and managers' involvement in the innovation process, which means that the top management takes care of employees and managers. Service is a production which requires the involvement of employees and managers in the innovation process because it is what could be called a "broad" organizational process. By this I mean that it is a process that involves the total organization and not only a small group of researchers. Involvement in service innovations is not an advantage, it is a must. Care, which means awareness of the employees' wellbeing and behavior and attempts to improve these, is, therefore, a prerequisite for service production.

"Service" is a broad category (see Illeris, 1996, for a definition) which is generally defined as the solving of problems that cannot be solved by the customer himself by use of a tool, a commodity. Services includes physical services such as cleaning, transport, operating hotels, knowledge services (for example education, consultancy, banks, real estate agency and personal services), health care services (for example hospitals), social services (for example social security and advice), psychotherapy and hairdressing to name but a few.

In the chapter I will, on the basis of earlier empirical studies, discuss theoretically how the innovation process in services can be conceived. First I will discuss the nature of innovation in services and present an overview of the literature on innovation in services. Then a model of the innovation process is introduced and it is discussed how care is a part of the process. Finally, I will provide three empirical examples to give a deeper understanding of how innovation with care is carried out.

INNOVATION IN SERVICES

What are innovations in services? I will start by discussing the nature of service innovation. There is a general overview of the literature on innovation in services which summarizes the research results and the general interpretation of what innovation in services is (Miles, 2004; Aa and Elfring, 2002; Gallouj, 2002; Sundbo, 1997, 1998; Boden and Miles, 2000; Coombs, 1999). The character of service innovations is discussed in this literature. A core element in the discussion has been whether innovation in services is different from that in manufacturing. The basis for this discussion are the many case studies that have been carried out about services firms (for example Miozzo and Soete, 2001; Sundbo, 1996, 1998; Boden and Miles, 2000; Howells, 2004; Vermeulen, 2001; Metcalfe and Miles, 2000; Andersen *et al.*, 2000; Sundbo *et al.*, 2001; Fuglsang, 2002; Gallouj, 2002; Brentani, 1993; Finch *et al.*, 1994). These case studies have been carried out in a variety of service industries (for example consultancy (van Poucke, 2004; Sundbo, 1998), engineering consultancy (Mattsson, 1994; Larsen, 2001; SIC, 1999), computer services (Jönsson, 1995), cleaning and other operational services (Djellal, 2002; Sundbo, 1999a) and tourism (Hjalager, 2002; Mattsson *et al.*, 2005; Sundbo *et al.*, 2007)). Surveys also play an important role. The well-known CIS-surveys (Community Innovation Survey) carried out by Eurostat (Innovation in Europe, 2004; Evangelista and Sirelli, 1998; den Hertog *et al.*, 2006; Drejer, 2004) have, since the early 1990s, included services. Other European surveys have also been carried out (INNO-Studies, 2004; Hipp and Grupp, 2005; Djellal and Gallouj, 2001; in Denmark SIC, 1999; Erhvervsministeriet, 2000).

It has been demonstrated that innovations in services are more complex and integrated. Namely, they are often product, process, organizational and market innovation in one and they are often small improvements (Voss *et al.*, 1992; Boden and Miles, 2000).

Innovations in services are rarely radical or large-scale (see Sundbo, 1998; Gallouj, 2002), but are mostly small improvements of products and procedures. Services are the delivery of a complex process in which the service is marketed contemporaneously with its production. Service innovations may be of different kinds. They may be product innovations (a new service product), process innovations (new procedures for producing the service), delivery innovations (new ways of delivering the service including peripheral service (see Normann, 1991)) and quality assurance (see Edvardsson *et al.*, 2000); market innovations (new behavior on the market or new strategic alliances), or organizational innovations (new organizational forms, for example new structures or a new organizational culture.

A service is fundamentally a behavioral act, and innovation in services a renewal of human behavior. This behavior often implies the use of technology, but the act is essential, which is why care is so important. This is also the reason why it is a service and not a need that can be satisfied by the customer buying a commodity. The service must be produced and delivered by a person at the moment of consumption. There are innovations in service technology; for example knowledge services (such as accountancy, consultancy, education) use IT and many services (for example insurance, banking) cannot be carried out without the use of IT to administer the service. All in all, innovations in services are both behavioral and technological, however, they are more behavioral than in manufacturing. Empirical investigations have shown that 16 percent of the innovations are technological and 30 percent depending on technology, and 54 percent of innovations are non-technological (Sundbo, 1998). Service innovations have increasingly become technological, particularly in knowledge services where the service can be delivered as a self service via IT-networks (the Internet, mobile telephones and so on). Thus, when we talk about service innovations, we may be talking of both technological and behavioral innovations and often a mixture of both.

Service innovations are often integrated, which means that they are product, process, organizational, delivery and market renewals at the same time. Even though they are integrated, they are often only small steps in renewing the services. Examples of typical service innovations could be a new insurance policy with different conditions and premiums; a cleaning contract not defined by the procedures but by the result (whether the customer is satisfied that things are clean), a new way of measuring employee satisfaction presented by a management consultancy, a hotel introducing free fruit in the reception. One may even discuss whether such small changes are innovations (see Sundbo and Gallouj, 2000). However, they develop the service firms and create economic growth. An attempt to create a distinction between innovation and the daily changes that everything goes through has been to argue that for something to be an innovation the change must be reproduced (Sundbo, 1997; Gallouj, 2002). The new service for a customer must be repeated for other customers, the new procedure or organization for producing and delivering the service must be widespread in the firm and so on.

Service innovations have also, in one tradition, been conceptualized as service development and seen as new solutions developed on the basis of the observation of service quality problems (Edvardsson *et al.*, 2000). Service innovation is described as service design and the tradition is oriented towards practical solutions (for example Gummesson, 1991). Thus, service innovations in this tradition are seen as a result of a kind of service engineering where new services are constructed, often with the aim of improving the service quality.

THE ORGANIZATION OF INNOVATION

Innovations in services are mostly behavioral. Thus, they are different from the mainstream product innovations in manufacturing, which are technological. However, there are similarities. Some service innovations are technological and some innovations in manufacturing are behavioral. But what about the innovation process – the way in which the work with innovation is organized? Is that different from what we know from the mainstream literature, which is based on manufacturing? The nature of the innovation process in services will be discussed in this section.

Innovations in services are not laboratory or science based (Sundbo, 1997; Gallouj, 2002) as they typically are in manufacturing. Often innovations are ad hoc, based on ideas from employees or managers. They may be part of a more systematic process, however the innovations are not top-down dictates, but must be developed along the way when ideas occur. Even when the organization attempts to have a very systematic innovation process, the concrete innovations must be developed by employees in the organization. This situation in particular requires that many employees and managers are not only involved in the innovation process, but also that they feel involved. Innovation in services is extremely dependent on the employees' and managers' commitment to innovation and care. This also applies to the top manager, who should be open to the active involvement of the employees.

Customers also play a central role in service firms' innovation processes. The philosophy of service production is – according to the service management and marketing theory (van Looy *et al.*, 1998; Grönroos, 2000) – based on the customer as a co-producer. All this means that innovation in services is a process which greatly involves different actors. It is not left to experts as is often the case in manufacturing where scientists make most of the effort.

Innovation has been described as a dual process that is both top-down and bottom-up (Sundbo, 1996). The employees and middle managers get ideas about new service products or new ways of producing and delivering the services and fight for realizing their ideas. Such processes appear naturally in all organizations. Often they die because the top management do not react to the ideas or even reject them and signal negative sanctions towards employees who use their time to get ideas and argue for them. This is not an optimal situation for the top management since innovations do not surface in such an atmosphere, or at least fewer do. The top management is rarely able to produce all the necessary ideas themselves and they do not normally have an expert apparatus with the particular task of producing innovations. The top management therefore becomes dependent on the employees presenting ideas and thus on a care system.

However, the number and type of ideas should also be limited, if not, the organization can waste many resources on getting ideas and communicating about their realization. The top management therefore continuously makes decisions concerning which ideas should be realized and which should be rejected. This places the top management in a power position, but it also forces the top manager(s) to be involved in the innovation processes.

Innovations often come from the top. The top managers get ideas or set up a framework for innovation activities. However, the top manager cannot develop and implement the innovations himself. He needs the employees in that process. Whatever the idea presented by the top manager or an employee, it leads to other employees and managers being involved in the process when the innovation is to be employed. A service is a behavioral activity that involves many people. Finally, the new service must be delivered by some service personnel. A service is not a commodity that can be produced by a machine and stored. It is an activity that must be produced by people at the moment of consumption and often the customer must be present. All this leads to the fact that innovation in services is a process that naturally involves many people and if the process is to be successful, all these people must feel involved and cooperate, which means care.

The development and implementation of innovations is normally organized as project work. If the management decides to continue with the idea, a project group representing different departments and often including the idea-maker is established. The project group develops the idea to a prototype that is taken over by the department which will produce the new service or implement the procedure. To have a successful development and implementation process requires that all these people feel involved.

The customers have a particular position in service production as mentioned. This is also the case concerning innovation. The service personnel pay much attention to the customers and how they perceive the service. This is both related to observing possible quality problems, how the service works for the customer – whether it solves his problem – and if he will buy the same or other services from this firm again. This service-attention also implies that employees often find a better solution for an individual customer; for example, a better procedure for cleaning or a more convenient type of pension saving. This better solution could be new and should then be diffused to other service workers in the firm (reproduced) and thus become an innovation. Often the service employee observes that the customer has an unsolved problem which could be solved by a new service. This is also the basis for an innovation if the employee presents it as such in the service firm.

Innovation in services is often customer based and customers are involved in the innovation process (for example Howells, 2004). This does not mean that the innovations are customer determined. The customers do not directly present ideas for innovations. Service firms that have tried to involve customers in this way have had bad experiences. Customers are rarely able to tell which new services they want in the future. Customer involvement is via interaction with employees, who then develop ideas for innovations. When the top management wants to involve customers in the innovation process, which is a good idea, it must nurse the employees' meeting with the customers (the "moment of truth" in service production, (Carlzon, 1987). Here there is a particular emphasis on the employees getting innovative ideas from this meeting. The customers again play a role in the later stage of the innovation process where a prototype of the innovation must be tested. The prototype is tested on customer groups, either in the form of discussions in focus groups or on a limited market segment.

Because innovations in services often are behavioral, there is normally no R&D or other expert department which carries it out, and the service firms are dependent on the employees involving themselves in the process. It is therefore difficult to set up an ideal model concerning how the innovation process should be organized. Improvements in innovation ability becomes a matter of organizational learning (Argyris and Schön, 1978; Nonaka and Takeuchi, 1995). Managers and employees learn how to organize the innovation process from earlier cases and from other firms.

The continuous learning process also requires care and involvement of the employees. These have the experience of what went well and what went wrong and they are those who will put the new forms of behavior into practice. Thus the management is dependent on their involvement in the improvement of the capability to innovate. The management itself, including the top management, must also be involved in the learning process, if not, this will not be taken seriously by the employees.

Entrepreneurship defined as establishing new firms on the basis of an innovation has not received much attention in relation to services. Only a few analyses can be found (for example Sundbo, 1998, and Morrison *et al.*, 1999 in tourism). However, most new firms are within services, however, these firms are very often not innovative and not growth-oriented. Nevertheless, entrepreneurship does count for an important part of service innovations.

In manufacturing it has been observed that innovation systems exist (Nelson, 1993). These are coherent networks of collaborating firms and institutions. In services, there is rarely a coherent innovation system. Many actors and streams of knowledge inputs (service professional, technological, management philosophical, trends in society and so on) are involved, but they do not act as a coherent system (Sundbo and Gallouj, 2000).

SYSTEMS OF INVOLVEMENT AND CARE

I will now present a model of the innovation process based on the research referred to above. The model consists of an overall conceptual framework based on the concept of strategic reflexivity, which will first be introduced and a phased categorization of the innovation process. The model divides the innovation process into three phases, the idea phase, the development phase and the implementation phase. The conceptual framework, it is argued from a statement of care, is a natural approach in service innovations.

The involvement of actors and how the innovation process can be considered as care-taking is discussed. I have stated that involvement of actors is important. However, the question arises: who are the actors in service innovations? Three main actors can be identified, namely managers, employees and customers. They have different roles in the three phases.

Service Innovation Processes and Care

Care is central in service innovation processes because these processes are based on the broad involvement of many employees, managers and customers and not only on expert-based R&D and technology development. The innovation processes are complex social processes. Care means that the top management is aware of the innovative potential of managers, employees and customers (and other external actors) and nurses these potentials, but also that it sets limits for intrapreneurship based on the strategy. Many service firms do this, and I will later give examples of how they do it, however, most service firms could improve their care for the innovation process.

The care aspect can be expressed in a conceptual model, which will be explained in the next paragraph.

A Concept for Careful Service Innovation – Strategic Reflexivity

Innovations in service firms must be understood as careful and coherent processes based on strategy forming a guideline (strategic innovation see Sundbo, 2001). Even though innovations often come from loose ideas through intrapreneurship, most service firms, at least the successful ones, have a general principle about which innovation to develop. A conceptual model for this principle, called strategic reflexivity, has been developed by Sundbo and Fuglsang (2002, 2006). This model will be explained in this section and related to actor involvement and care. Strategic reflexivity involves managers and employees actively, not only in the detailed innovation processes, but also in the more general strategy process. Strategic reflexivity is a general principle that may be found in all organizations no matter which sector or industry, but it is particularly central in service firms.

The model sees the innovation processes in service firms as a broad coherent process. Innovation is about the future of the firm. The service firm is developed within a broad framework which relates to the goals for where this firm will be in the market in the future. These goals and the overall means for getting there are expressed in the strategy as explained earlier. The strategy is the general guiding principle for the development of the firm. Most service firms introduce service concepts, which are broad service fields with a certain business idea. A service concept could, for example, be SAS's business traveler's airline in the 1980s where SAS wanted to serve the business traveler. This included high quality service counting peripheral ones such as admission to SAS hotels, thus the journey could be totally booked via SAS. The price was very high. Another concept in airline services is the more recent concept of discount flying that low-price airlines such as Ryan Air, and Easyjet have introduced. In that concept there is no extra or added services such as meals, hotels and so forth, and even the core service may be of poor quality (delayed planes and so on). However, the price is very low. Such broader concepts are not only a product, they are a bunch of products, processes, organizational forms, culture and attitudes. The strategy specifies in general terms which type of service concepts the firm should have. Innovations are the details that fulfill the concept. The concepts are defined by these details. Therefore innovations in services are small renewals which together change the concept.

The strategy is an interpretation of the future market (see Mintzberg, 1994). As nobody can know the future market, the service firm can only provide a qualified guess and act on the basis of it. In the process of formulating the strategy, many service firms involve the employees and managers in order to have the broadest range of experiences, analyses and ideas. The top management makes the final decision concerning the strategy, but the employees and middle managers contribute to the process and can influence the strategy.

When the strategy is decided, it must be implemented and work. The strategy defines the kind of service concepts (for example "are we a business travelers' airline" or "discount airline") and guides the innovation processes (define areas for possible innovations and the borders outside which innovative ideas can not be accepted). However, since the strategy is based on an interpretation of an unknown future, it is important to follow the market and general societal trends to see if the pre-assumptions and the implementation of the strategy is still valid. This requires involvement of employees and managers. They follow the market, among others through direct customer contact and they have a broad involvement in society. The employee and managers should, therefore, reflect on the strategy and whether it still works. Therefore the model is called strategic reflexivity.

This reflexivity also includes considerations about whether the types of innovation and the concrete innovations are the right ones. All employees and managers should be potentially involved in these deliberations. Some have this as a working task, the others are to present their considerations if they discover a problem. Such problems may include that the strategy does not seem to work, the service concept does not seem to work or the innovations that have been implemented have not been sufficiently successful. The reflections may be based on systematic analyses and market investigations, on knowledge about new technology or service principles that the firm has not considered or just on concrete daily observations.

Changes may also come from employees or managers acting as intrapreneurs by getting ideas for innovations and fighting for having them accepted. The new ideas may be within the framework of the strategy and thus the intrapreneurship is a contribution to an optimization of the strategic goals through involvement. They may also break the strategic framework. In most such cases they will be rejected. However, in some cases reflections within the management or among the employees may lead to one innovative change of idea – not only the service concept, but the whole strategy.

The reflections and innovative ideas and intrapreneurship thus may lead to an adjustment of the principles for which types of innovations should be developed, to the introduction of new types of service concepts or to an adjustment of the strategy. In rare cases it may even lead to the introduction of a new strategy. These decisions are made by the top management, but the employees and middle managers are involved in the process through their reflexive roles.

The strategic reflexive innovation mode also includes the creation of the different in the innovation processes that have been mentioned. These roles can be institutionalized. The management may be aware that there are people who play all the necessary roles and, if not, they can employ such people or encourage such behavior to be developed among the employees. This is a core part of care for the innovation process.

Customers do have a central role in the model of strategic reflexivity, not as directly involved actors. They are the final judges that decide on strategic and innovative success, but they can rarely say beforehand whether they will accept an innovation or not. They are actors in which employees and managers of the service firms reflect the service concept and observe the reaction – a mirror for the strategic reflection process. The employees and managers are the active partners in this mirroring process. The customers may have relevant reflections, which may be the basis for strategic and innovative adjustments. The service firm must, however, organize the communication channels for these reflections. This is done for example through complaints departments or hotlines for IT-network services. Through complaints and customer problems presented to the departments many service firms get ideas for innovations and even strategic adjustments. They are in many service firms the most important sources of innovation (see Edvardsson *et al.*, 2000).

The Idea Phase – Innovation as Strategic Innovation

Managers, employees and customers are the most important idea-makers of all. Ideas do not come from nothing, but from inspiration, from sources of information or other people. Ideas may come from each of the three main actors, but as we have already discussed, they often come from interaction where several of these actors are involved. We have mentioned the interaction between employees and customers. However, the interaction between employees or managers and between employees and managers also leads to new ideas. However, it is not sufficient that the employees and middle managers get new ideas. They need to communicate the ideas throughout the firm and convince others including the top management (or the owner) about the idea and struggle for realizing it. Innovation in service firms is often based on corporate entrepreneurship (*Strategic Management Journal*, 1990) or intrapreneurship (Pinchot, 1985). The firm must be in a situation where the promotion of ideas is encouraged. This may be done by producing a general creative climate (see Ekvall, 1996) by including innovation in the organizational culture (see Schein, 1992). Ideally, there must also be a formal system for presenting ideas, particularly if the firm is large. This might be boxes in which the employees can drop new ideas or a committee to which they can send the ideas. It may also be some managers or the top management itself that can be the address to which to communicate new ideas. It is extremely important for maintaining employee involvement and care that the idea-presenting process is treated in the right way. This means that offering ideas should be accepted and encouraged; all ideas should be considered and the creator receive an answer. The management, who makes the decision, do not need to accept all ideas to maintain a high level of involvement. Employees accept that their ideas may not be introduced if they get a reason. The employees should also be encouraged to pay attention to customers' problems and need to get new ideas for innovation. There are examples of service firms (for example the cleaning firm ISS; Sundbo 1999a; SIC 1999) which have given the front personnel (for example cleaning assistants) power to sell new services to customers. Most important, a fact which many service firms forget, is that one good service solution for one customer invented by one employee should be communicated to the rest of the organization thus it could become a real innovation (as we have defined it: a reproduced solution).

Care is carried out as a role system, that is different social-psychological roles that employees and managers can play. Altogether the roles constitute an organizational innovation system because each role plays a part in the activities necessary to create innovations. The idea phase is characterized by intrapreneurship, however it is rarely one person who carries out the whole process from idea to implementation. There are different roles (for example social psychology, Mead, 1934). Idea-making and entrepreneurship is one role (see Sundbo, 2001; Lessem, 1984). Another is the knowledge-gatekeeper or analytical role; this is the person who puts the idea in an analytical framework – what do we know and will there be a market? A third role is the interactive one; this is the person who can convince others in the organization about the idea. Sponsorship or championship (Pinchot, 1985) is also a role; this is another person, often a senior one, who supports the idea-maker. Decision-making is a role. These roles are some that the employees and managers can play in addition to their normal position. They are professional roles, often institutionalized by the management, however, they often engage the person beyond their working hours and usual effort. One person may play several roles and the roles may be temporary. By this professionalization of role-playing, the employees avoid becoming too involved and going beyond their work obligations.

Also top-down innovation processes involve customers in the idea phase if the service firm is to be successful. If the top management starts an innovation project, it should involve customers in the assessment of whether there will be a market for it, how it should be presented to the market an so on. Many service firms co-operate directly in a common project with customers in developing innovations. This is, for example, the case when ISS' food service (cleaning in the food industry) develops a new cleaning procedure in co-operation with a slaughterhouse or when a consultancy firm develops a new personnel management system together with a client. Often it is difficult to distinguish between whether the innovation has been developed by the service firm or the business customer and which of them own it. Private customers are rarely involved in this way. They may be involved in a kind of permanent focus group that follows the innovation process.

Other actors may also be involved in the idea phase. Suppliers are often involved and sometimes they make the innovations. For example, suppliers of ATMs have made innovations for the banking sector (Sundbo, 1998) and suppliers of kitchen technology for the restaurant sector (Hjalager, 2002). Service supplier firms also make innovations for other service firms, for example when consultancy firms develop new procedures and organizational patterns or new marketing for service firms. Suppliers can also function as idea-making partners in interaction processes with employees or managers. Service firms also participate in innovation networks with other actors – competitors, firms from other sectors, public institutions and so on – where ideas for innovations may occur. Service firms participate in such networks more than manufacturing firms (Innovation in Europe, 2004), even with competitors, which is surprising due to the fact that service innovations are normally very easy to imitate. Further, competitors function as a source of ideas for innovation because service firms often try to imitate their competitors and competitors' moves in the market are often the most important reason for innovating (Sundbo, 1998). However, one may hardly characterize the latter as the involvement of competitors.

Despite the requirement of large employee, manager and customer involvement, the service firms also restrict the idea-making and innovation activities and the involvement (Sundbo, 1998). Care is also to know when to limit the free display of entrepreneurship. This – what we have termed "broad" – innovation process, which involves maybe most employees and managers may result in that many resources (up till more than 20 percent of the working time, see Sundbo, 1998) are used on innovation activities. This may drain the resources from production and selling activities and thus imply as risk of suboptimal production capacity. The ideas may also lead the firm in different directions thus there is no common direction for the development. To restrict the suboptimal use of time and to give the firm's development a direction, the top management regulates the idea-making and the total innovation process. First, it sets up a direction, which is expressed in the strategy. The strategy does not need to be a very formally written and analytically based plan, but may be just a loose business idea in the mind of the owner. Whatever, it gives a direction that the firm should follow – at least until a new fundamental strategy is introduced. This means that the top management and other managers have a framework for deciding which ideas to continue with. The strategy can also be an inspiration for innovations within the chosen framework. It should thus be careful and involvement oriented, and the involvement of employees and managers and their interaction with customers is explicitly expressed in the strategy of successful service firms. The innovation system in services may thus be characterized as strategic innovation (Sundbo, 2001).

The Development Phase – Project Organization

When an idea is accepted in the service firm, a development process starts. The idea must be developed and become usable. This can be done in many ways, but mostly it is done by establishing a project group that has the task of developing the idea. Generally, several departments and functions are represented in the group. In particular, the marketing department has a central role as the market possibilities are the primary criteria in success assessment, but normally different operational (production) departments are also represented. To motivate an employee to involve him or her in innovation processes and as a reward to the idea-maker, he or she is often included in the project group. The idea-maker may even be leader of the group. The idea-maker sometimes follows the idea the whole way through to full implementation. The idea-maker may get their own implementation team and later production. Many intrapreneurs state that the reward for the involvement they are looking for is independence, which they will get if they get their own group or department, or at least an independent position to work within development and with the implementation of the idea. Thus, sometimes, the development is left to the intrapreneur.

The development phase does not require that much involvement, it is mostly an introverted working phase where the project group procures knowledge, solves problems and assesses market possibilities. In relation to the latter, the marketing department must be involved as must the top management, at least enough to support the work in the project group and substantiate decisions. Decisions may include a closure of the project. The top management follows the development process and makes regular decisions concerning whether the project should continue or not depending on the chances for market or implementation success (if it, for example, is an organizational innovation). To ensure a general positive attitude, employees and managers have to involve themselves in idea-making and development activities such as membership of a development project group, it is important that the top management substantiate its decisions and interact with the project group. Many project group members do this work besides their normal duties and thus make an extra effort, sometimes in their spare time.

The project group members are often involved. This depends primarily on the project management (for example Cooper, 1998). The project manager and all project members should be aware that a common effort is the condition for success. Care becomes, in this phase, a common responsibility. If it is a good project group, the project members have different competencies, experiences and personal networks. This variation can be utilized, thus the development process includes different aspects and views to ensure a good and successful result. Even though the other internal actors are not directly involved in the development process, the project group needs information and at least a limited effort from them. The project group also needs acceptance of the project from the other actors, which also could include customers and public regulation authorities, to ensure an efficient development process and implementation in the future. In project groups different roles must be represented to ensure a successful process. This may, for example, be the role of idea-making, the skeptical role, the communicative role and so forth. In research, four particularly important roles have been discovered: the gardener (creates well-being and an innovative climate), the conceptualizer (clarifies concepts, secures agreement), the jester (stimulates ideas and asks "dumb" questions) and the challenger (screens the group's knowledge, gatekeeper of facts) (Darsø, 2001).

It is also important that the project group members are motivated to learn from the development process and that the firm has a procedure for storing the experiences, both from successes and fiascos (a care-learning system). Since the innovation process in services cannot be put on a standardized form, it will have new aspects each time a project group is established. The project groups cannot just follow a prescription, but must, to some degree, make their process. However, they can benefit from experiences had by other project groups. They should have access to earlier experiences, both in the form of tacit and more explicitly formulated knowledge (see Nonaka and Takeuchi, 1995).

The Implementation Phase – Accept on the Market and in the Organization

When the service innovation has been developed, it must be implemented. If it is a new service product or a new form of market behavior, it must be launched on the market. If it is a new production or delivery process or a new organizational form, it must be introduced into the firm. In this situation involvement and care are also necessary. The engagement of the actors does not need to be very high because there are procedures for implementation and a department that takes care of it. It is no longer the engagement and commitment of single individuals that takes the innovation process forward. However, the implementation phase requires acceptance by the employees and managers, irrespective of whether it is a new procedure or a product. New processes and organizational forms require acceptance to work in praxis. Organizations often resist change (Sundbo *et al.*, 2001) and this opposition must be overcome if implementation is to be successful. The management should therefore involve the employees and middle managers to the extent that they accept the innovation or at least not actively resist it.

If the innovation is a new service product, it is important that the employees understand and accept the product. The new service must be of good quality when delivered to the customers, which means that the customers feel it solves their problem and there is no failure in the delivery process (Edvardsson *et al.*, 2000). Care for customers is very important in services. The new service must be produced and delivered by some employees. If they do not understand and accept the new service, there will be too many quality problems. They must therefore be involved in the implementation, thus they can influence the new service and the implementation process. They have the general experiences with quality problems and can contribute to making the new service better and more successful on the market. Other employees may have back-office back-up functions for the new service. These employees and their managers also must be involved in the implementation process and be positive so as to correct eventual quality failures in the internal procedures.

Most important is that the market accepts the innovation. This is obvious when it concerns a new service product, delivery innovation or a market behavior. However, even when it is a new procedure or organizational form, implementation may influence the quality of the services and the way the customers perceive the service. Customers are therefore the most important actor in the implementation process. It is difficult to involve the customers to make them feel positively engaged in the implementation process. Customers are mostly distant neutral actors in the market. The service firm may attempt to involve selected groups of customers in test groups or the like. It may also attempt to create a broad customer loyalty on the market. The means could for example be customer clubs where customers could exchange opinions about the innovation, or an image creation in the public, for example by story-telling about the firm and the innovation. The launching of the innovation on the market may also be followed by an event such as a TV-show or a concert. Such events may, in the experience economy (Pine and Gilmore, 1999), create some customer involvement in the service. These active forms of marketing can mostly involve customers, both as potential purchasers and as active contributors to the implementation process; they may put forward proposals for the improvement of the innovation and how it could be launched better. The firm may also do market surveys throughout the implementation process to measure whether the market accepts the innovation and many service firms do that. However, even though this is a necessary control tool, it does not involve customers in an active and positive way and thus does not push the implementation process forward. It is more an instrument for the management to decide whether the implementation process should be stopped. More active customer methods such as focus group interviews, adding experience and story-telling elements, user-clubs and so on are needed.

EMPIRICAL EXAMPLES

In this section I will give four different examples of care-based innovation processes and systems in service firms. They will emphasize the involvement of employees and organizational learning, which is also a part of the care-system. These practical examples can give a deeper understanding of the service innovation process, thus supplementing the theoretical discussions that have been presented in this chapter. The different cases present different ways in which the firms take care of the innovation process and involve actors (primarily the employees) in the process.

Roles in Innovation Processes in an Insurance and a Payment Company

An insurance company and a payment company (sells credit and payment cards and administers payment transaction via cards) have the same type of innovative organization (Sundbo, 1998). This organization follows the principle of a dual organization. Innovation activities were encouraged, but the management also rejects many ideas and of those accepted many are abandoned by the management throughout the development process.

The study observed that the employees and managers play different roles in the innovation process. Three roles were identified, the strategist, who reflects, analyses and gives suggestions for new strategy, the intrapreneur who acts and fights for realizing innovative ideas, and the producer who does the daily routine job whatever that may be.

These cases show how innovation with care can be carried out in a balanced way that both calls for creativity and the active effort of the individuals and restricts the use of resources thus the costs are controlled. There is much innovative activity in both firms. Managers used about 30 percent of their time on innovation (entrepreneurial) activities, employees used about 15 percent. Managers used about 25 percent on strategy and employees used about 10 percent in one company and 20 percent in the other. However, the innovation activity is disciplined. The top management decides regularly upon which ideas and projects should continue. The employees are not allowed to carry out the intrapreneurial role as much as they might wish. The desire for playing an intrepreneurial role and the actual role-playing was not correlated at all. Those that desire to play the role were those at the top and the bottom of the hierarchy: older people with a high position and high seniority and young persons at the lower end of the hierarchy with low seniority. It was mostly the first group that was allowed to carry out innovative activities. Further, the intrapreneurship was not a function of the corporate culture as one might have expected (that is the more creative and change oriented culture, the more intrapreneurship). The strategy and management-decision system decided how much intrapreneurship arose.

This means that these two companies have an unrealized innovation potential in the employees and may have missed opportunities for successful innovation. The care is not optimal. On the other hand, many of these potential innovations are outside the chosen strategy and thus eliminated by the management to maintain a red line in the development.

Customer Orientation and Employee Empowerment in a Bank

A small Danish bank has created a program that should make the employees more customer oriented and involve the employees in the innovation process and thereby create a learning organization (Sundbo, 1999b). This is an example of a very conscious and planned care approach to innovation.

The bank wanted to achieve its aims by increasing the employees' competencies. The means were four development projects. One was called "the learning organization." All managers at all levels were given a course about the learning organization, communication and team training. They were to go back to their employees and spread the knowledge and involve them in the development of the learning organization. Another project was a group-oriented innovation project. All employees could send ideas for innovation to the management. If the management accepted the idea, the employee could establish a development group involving some of his or her colleagues that he could select himself. They could develop the idea. If the management decided to implement the innovation, a new implementation group was established of which the idea-maker became a member. The third project had the general aim of involving employees and managers in the strategy process, innovation activities and modifying their daily work in the treatment of customers, team work, use of technology and so on. All employees and managers were at courses which included practical exercises. The fourth project was a course in personal development and team building. The course included practical group work where the groups should develop a new bank product.

Innovation is a factor in firm development and growth. In this service firm, innovation was not seen as isolated as a kind of free-standing R&D project. Innovation was seen as integrated with the strategy and the process of formulating and developing the strategy was also seen as integrated into day-to-day work such as improving perceived customer quality (Edvardsson *et al.*, 2000), efficiency and flexibility (the employees could help each other instead of defining their own work area and only concentrate on that). Innovation thus is a "broad" activity in the sense that it involves and is integrated in many functions (strategy, quality assurance and so on). Even though there were concrete innovation projects that could be clearly identified, the origin of these projects and the decisions about whether to continue with the ideas were integrated in development of the strategy and improving perceived customer quality. Innovation is also a "broad" activity in the sense that it involves the whole organization. The whole program had the purpose of involving the employees in the innovation process and creating intrapreneurship, thus the employees themselves took initiatives to start innovation processes.

The program resulted in several concrete innovations and the employees became, generally speaking, empowered, that is they took independent initiatives, became more flexible, dared to be more active in customer relations and so on. The organization also became more learning oriented in its approach. The employees became more learning oriented, however the problem was to store the experiences, so that they could be transmitted to future employees and managers. It was attempted to solve the latter by creating tacit knowledge and making IT-based experience files. This part was not as successful, as the employee empowerment and innovation parts of the project. People later forgot to look in the IT-files and to inform new employees about the tacit knowledge. Further, not so many new intrapreneurs were created. The employees were more aware of the innovation requirements and became more willing to participate in the innovation processes, but as role players in parts of the process. Not many came out of the process as personalities who wanted to fight for their innovative ideas and carry them the whole way through to implementation.

This case demonstrates that innovation with care can mobilize unused resources in the firm. However, it also demonstrates that even when the management attempts to practice innovation with care, it is difficult to mobilize long-term intrapreneurship where employees persistently try to realize their idea and learn from the intrapreneurial processes.

Change of Employee Behavior in a Hotel

A middle-sized hotel is a typical tourism firm. The problems for tourism firms are among others that the employees are not well-educated and the personnel turnover is high, thus it can be difficult to assure a high level of customer-service. Further, the hotel product is fairly standard, thus it is difficult to differentiate from competitors. Consequently, the hotel branch is a very competitive one where it is difficult to increase profit. Many hotels try to differentiate themselves by low prices or high standards (many facilities and amenities in the rooms etc., which leads to more stars in the classification system). The latter raises the demands in relation to employee competencies, which the lack of tradition for stable professional staff make difficult to fulfill. This hotel decided to differentiate itself by becoming a modern design hotel with artistic drawings in the rooms, modern furniture and so on. It then became oriented towards a certain market segment, a kind of younger people from the "creative class" (see Florida, 2002). The manager acted from a perspective of care for customers in the form of attracting a particular customer segment and giving them a high level of service, and care for employees because they were to deliver this service. The employee-customer interaction thus is the basis for development of the hotel and thus innovations.

This implied that the staff should introduce a new behavior towards the customers. They should be more modern in their attitude, their language and the way in which they solve customer problems. They should all in all be more service oriented without the costs. Employee behavior was a part of the innovation – the modern design style. It was, however, difficult to change the behavior and attitude of the employees from the top. Therefore the manager of the hotel decided to involve and empower the employees.

He started a project called "self control". The employees were involved in the implementation of the new strategy. Even though this was decided from the top, the employees got an insight in the idea of the strategy (thus it was not just one of these "troublesome new things the management has invented"), and they learned from each other about how to behave towards the guests. The employees were organized in self-governing groups that regularly used a part of their working time to discuss their work and behavior and to exchange experiences. These were learning groups which carried out organizational learning in practice.

This is an example of how employee involvement is a necessary part of the innovation (the product) in services besides an advantage in the development process that leads to implementation of the innovation. Employee behavior and even attitude is a part of the new product, which can be better characterized as a new concept because it implies a new strategy, a new market segment, new physical artifacts and new customer-behavior. The employees' interaction with customers and customer care was an important part of the innovation-with-care system, which was successful in introducing several innovations. However, these innovations were small, incremental ones, which suggests that innovation-with-care does not necessarily lead to radical innovations, but may be a good approach to developing incremental innovations and improvements of existing products and routines.

CONCLUSION

Innovation in services is a strategic reflexive process in which care is important. Care is executed as a management and organizational principle in the cases that have been presented.

Innovation processes in services are generally characterized by a dual organization of innovation: Innovation may come top-down, but mostly it comes from the bottom upwards. The top management steers the process. Innovation in services is particularly dependent on many employees' and managers' involvement. This involvement is first and foremost expressed as corporate entrepreneurship or intrapreneurship. Innovation is strategic, it is carried out within a strategic framework, which is the guide for the firms' development and decisions and thus for the kind of innovations that are wanted and accepted. An intrapreneur may in rare cases break the strategy and thus introduce a new strategy by carrying through a radical innovation. Innovation processes are improved through organizational learning. The involvement of employees and managers is often made via different roles that are limited and professional. Even though innovation still requires engagement beyond normal work duties, the role-making reduces the personal involvement. Via awareness of the different innovation roles, the management can create an institutionalized system of roles that can ensure a successful innovative organization.

Care is also extended to external actors such as customers. The hotel and bank case showed that customers were not directly involved in the innovation process as providers of ideas for innovation, but the employees' careful interaction with customers led to new innovative ideas.

External actors have not traditionally been as involved in services as they, for example, have been observed to be in manufacturing (for example Håkansson, 1987). This is partly caused by service innovations being extremely easy to imitate and a tradition of not co-operating with competitors has developed in many service industries. However, this situation has changed so service firms now involve external actors more than in manufacturing (Innovation in Europe, 2004). Customers have a central role in services as they are supposed to participate directly in the production process (see the service management theory, Grönroos, 2000). Customers, and customer-care, therefore, also have a particularly central role in the innovation process in services, not as directly active partners, but as mirroring "others" (see Mead, 1934). The employees get new ideas for service innovations that can solve customer problems from the interaction with customers. Customers are also involved as test-mirrors through the innovation process.

To conclude, service innovations are based on care for the strategic reflexive processes and the actors and roles involved. This care does not only mean nursing of the actors and encouraging intrapreneurship, but also restrictions and a strict decision making processes concerning new ideas and innovation projects. The optimal care-restriction balance is difficult to find in practice as the insurance and payment-company cases demonstrated.

REFERENCES

Aa, W. Van der and T. Elfring (2002), "Realizing innovation in services," *Scandinavian Journal of Management*, **18** (2), 155–71.

Andersen, B., J. Howells, R. Hull, I. Miles and J. Roberts (eds) (2000), *Knowledge and Innovation in the Service Economy*, Cheltenham, UK and Northampton, MA, USA: Edward Elgar.

Argyris, C. and D. Schön (1978), *Organizational Learning: A Theory of Action Perspective*, Reading, MA: Addison-Wesley.

Boden, M. and I. Miles (eds) (2000), *Services and the Knowledge-Based Economy*, London: Continuum.

Brentani, U. (1993), "The new product process in financial services: strategy for success," *International Journal of Bank Marketing*, **11** (3), 15–22.

Carlzon, J. (1987), *Moments of Truth*, Pensacola, FL: Ballinger.

Coombs, R. (1999), *Innovation in Services*, Nijmegen lectures on innovation, Antwerpen: Maklu.

Cooper, R. (1988), *Winning at New Products*, Ontario: Addison-Wesley.

Darsø, L. (2001), *Innovation in the Making*, Copenhagen: Samfundslitteratur.

Den Hertog, P. (2000), "Knowledge-intensive business services as co-producers of innovation," *International Journal of Innovation Management*, **4** (4), 491–528.

Den Hertog, P., T. Poot and G. Meinen (2006), "Towards a better measurement of the soft side of innovation," in: J. Sundbo, A. Gallina, G. Serin and J. Davis (eds), *Contemporary Management of Innovation*, New York: Palgrave Macmillan.

Djellal, F. (2002), "Innovation trajectories in the cleaning industry," *New Technology and Employment*, **17** (2), 119–31.

Djellal, F. and F. Gallouj (2001), "Innovation in services, patterns of innovation organisation in service firms: postal survey results and theoretical models," *Science and Public Policy*, **28** (1), 57–67.

Drejer, I. (2004), "Identifying innovation in survey of services: a Schumpeterian perspective," *Research Policy*, **33** (3), 551–62.

Edvardsson, B., A. Gustafsson, M.D. Johnson and B. Sandén (2000), *New Service Development and Innovation in the New Economy*, Lund: Studentlitteratur.

Ekvall, L. (1996), "Organizational climate for creativity and innovation," *European Journal of Work and Organizational Psychology*, **5** (1), 105–23.

Erhvervsministeriet (2000), *Service i forandring* (Services in Change), Copenhagen: Erhvervsministeriet/Ministry of Industry.

Evangelista, R. and G. Sirilli (1998), "Innovation in the service sector: results from the Italian statistical survey," *Technological Forecasting and Social Change*, **58** (3), 251–69.

Finch, R., J. Fleck, R. Procter, H. Scarbrough, M. Tierney and R. Williams (1994), *Expertise and Innovation*, Oxford: Clarendon.

Florida, R. (2002), *The Rise of the Creative Class*, New York, Basic Books.
Fuglsang, L. (2002), "Systems of innovation in social services," in: J. Sundbo and L. Fuglsang (eds), *Innovation as Strategic Reflexivity*, London: Routledge.
Gallouj, F. (2002), *Innovation in the Service Economy*, Cheltenham, UK and Northampton, MA, USA: Edward Elgar.
Grönroos, C. (2000), *Service Management and Marketing*, 2nd edn, Chichester: Wiley.
Gummesson, E. (1991), *Kvalitetsstyring i tjänste och serviceverksamheter* (Quality Control in Service Firms), Karlstad: Center for Service Research.
Hipp, C. and H. Grupp, (2005), "Innovation in the service sector," *Research Policy*, **34** (4), 517–35.
Hjalager, A.M. (2002), "Repairing innovation defectiveness in tourism," *Tourism Management*, **23**, 465–74.
Howells, J. (2004), "Innovation, consumption and services: encapsulation and the combinatorial role of services", *The Service Industries Journal*, **24** (1), 19–36.
Håkansson, H. (ed.) (1987), *Industrial Technology Development. A Network Approach*, London: Routledge.
Illeris, S. (1996), *The Service Economy*, Chichester: Wiley.
INNO-Studies (2004), *Innovation in Servcies: Issues at Stake and Trends*, INNO-Studies 2001: Lot 3 (ENTR-C/2001), Brussels: the EU Commisssion.
Innovation in Europe (2004), Brussels: EU Commission/Eurostat.
Jönsson, M. (1995), *Utveckling av konsulttjänster* (Development of Consultant Services), Karlstad: Center for Service Research.
Larsen, J.N. (2001), "Knowledge, human resources and social practice: the knowledge-intensive business service firm as a distributed knowledge system," *The Service Industries Journal*, **21** (1), 81–102.
Lessem, R. (1984), "A theory for corporative innovation and renewal," *Journal of General Management*, **9** (4), 17–35.
Mattsson, J. (1994), "Quality blueprints of internal producer services," *International Journal of Service Industry Management*, **4** (4), 66–80.
Mattsson, J., C.F. Jensen and J. Sundbo (2005), "An attractor-based innovation system in tourism: the crucial function of the scene-taker," *Industry and Innovation*, **12** (3), 357–81.
Mead, G.H. (1934), *Mind, Self and Society*, Chicago, IL: University of Chicago Press.
Metcalfe, J.S. and I. Miles (eds) (2000), *Innovation Systems in the Service Economy*, Norwell, MA: Kluwer.
Miles, I. (2004), "Innovation in services," in J. Fagerberg, D. Mowery and R. Nelson (eds), *The Oxford Handbook of Innovation*, Oxford: Oxford University Press.
Mintzberg, H. (1994), *The Rise and Fall of Strategic Planning*, New York: Free Press.
Miozzo, M. and L. Soete (2001), "Internationalization of services: a technological perspective," *Technological Forecasting and Social Change*, **67** (7), 159–85.
Morrison, A., M. Rimmington and C. Williams (1999), *Entrepreneurship in the Hospitality, Tourism and Leisure Industries*, Oxford: Butterworth-Heinemann.
Nelson, R. (1993), *National Innovation System*, Oxford: Oxford University Press.
Nonaka, I. and H. Takeuchi (1995), *The Knowledge-Creating Company*, Oxford: Oxford University Press.
Normann, R. (1991), *Service Management*, 2nd edn, Chichester: Wiley.
Pinchot, G. (1985), *Intrapreneuring*, New York: Harper & Row.

Pine, B.J. and J.H. Gilmore (1999), *The Experience Economy*, Boston, MA: Harvard Business School Press.

Schein, E. (1992), *Organizational Culture and Leadership*, 2nd edn, San Francisco, CA: Jossey-Bass.

Schendel, D. and D. Channon (eds) (1990), "Special issue on corporate entrepreneurship," *Strategic Management Journal*, **11** (Summer).

Service development, Internationalisation and Competence development (SIC) (1999), "Danish service firms' innovation activities and use of ICT, based on a survey," report no. 2, Roskilde University, Centre of Service Studies.

Sundbo, J. (1996), "Balancing empowerment," *Technovation*, **16** (8), 397–409.

Sundbo, J. (1997), "Management of innovation in services," *The Service Industries Journal*, **17** (3), 432–55.

Sundbo, J. (1998), *The Organisation of Innovation in Services*, Copenhagen: Roskilde University Press.

Sundbo, J. (1999a), "The Mannual Service Squeeze," Roskilde University, Centre of Service Studies, research report 99:2, accessed at www.ruc.dk/ssc/forskning/centre/css.

Sundbo, J. (1999b), "Empowerment of employees in small and medium-sized service firms," *Employee Relations*, **21** (2), 105–27.

Sundbo, J. (2001), *The Strategic Management of Innovation*, Cheltenham, UK and Northampton, MA, USA: Edward Elgar.

Sundbo, J. and L. Fuglsang (eds) (2002), *Innovation as Strategic Reflexivity*, London: Routledge.

Sundbo, J. and L. Fuglsang (2006), "Strategic reflexivity as a framework for understanding development in modern firms," in J. Sundbo, A. Gallina, G. Serin and J. Davis (eds), *Contemporary Management of Innovation*, New York: Palgrave Macmillan.

Sundbo, J. and F. Gallouj (2000), "Innovation as a loosely coupled system in services," *International Journal of Services Technology and Management*, **1** (1), 15–36.

Sundbo, J., L. Fuglsang and J.N. Larsen (2001), *Innovation med omtanke*, Århus: Systime.

Sundbo, J., F. Orfila-Sintes and F. Sorensen (2007), "The innovative behaviour of tourism firms: comparative studies of Denmark and Spain," *Research Policy*, **36** (1), 88–106.

Van Looy, B., R. Dierdonck and P. Gemmel (1998), *Services Management*, London: Pitman.

Van Poucke, A. (2004), "Towards radical innovation in knowledge-intensive service firms," PhD thesis for Erasmus University, Rotterdam.

Vermeulen, P. (2001), *Organizing Product Innovation in Financial Services*, Nijmegen: Nijmegen University Press.

Voss, C., R. Johnston, R. Silvestro, L. Fitzgerald and T. Brignall (1992), "Measurement of innovation and design performance in services," *Design Management Journal*, Winter, 40–46.

3. Customer Relationship Management (CRM) as innovation: taking care of the right customers

Jan Mattsson

Relationship marketing has become the dominant paradigm in marketing during recent decades (Sheth and Parvatiyar, 1995). Marketing will need to undergo great changes in order to find its functional (Doyle, 1995) and ethical role (Mattsson and Rendtorff, 2005) in the new millennium. Customer Relationship Management, or CRM, can be seen as a tangible response to the strategic challenges facing marketing (Mattsson et al., 2006). CRM is the idea of relationship marketing put into practical operation with reference to customers (Gummesson, 2004). Zeng et al. (2003) make a review of the key ideas and concepts in B2B and CRM literatures and distinguish between the two literatures, while underlining the importance of using strategies for integrating business processes which will benefit involved commercial partners.

However, it has been argued that CRM is more than a marketing process as it should be used as a strategic lever to support the company mission with the aim of becoming more customer-centric (Buttle, 2005). Recently, the idea of a strategic CRM (or Global CRM) has been proposed as encompassing the entire organization woven around the concept of customer value (Kumar and Reinartz, 2006). Jain and Singh (2002) underline that the use of such a kind of CRM necessitates a restructuring of the firm from being product-centric to becoming customer-centric.

CRM success depends on applying a cross-functional, holistic, and customer-centric approach simultaneously (Bull, 2003) combining people, processes and technology with the aim of better understanding and care for customers (Chen and Popovich, 2003). However, there seems to be asymmetric outcome effects of relationship strategy implementation. Excellent execution increased customer satisfaction compared to having no relationship strategy. On the contrary, poor execution led to negative effects exceeding the positive effects from CRM initiation (Colgate and Danaher, 2000). Bull (2003) has analysed such a case. In the digital environment, CRM has been related to satisfaction with web site, but not to increased sales or profit for the business as such (Feinberg and Kadam, 2002). All in all, academic research regarding the real impact of CRM on firm performance is rather limited (Buttle, 2005).

In this chapter the aim is to illustrate managerial experiences with the implementation of CRM systems in one large multinational industrial group and in a small manufacturing company in the same group. Following Mattsson *et al.* (2005) and Mattsson and Orfila-Sintes (2007), we take a broad definition of innovation comprising technical, as well as social innovation (Johannisson, 1987). The latter category may for instance include organizational change (Haglund *et al.*, 1995), change in service products and customer behaviour (Sundbo, 1998), market behaviour or other strategic changes (Sundbo *et al.*, 2001).

The term innovation will now be used to designate the innovation activities of implementing CRM.

Hence, we will view this implementation process of CRM systems as a special kind of organisational innovation directed towards customers. As new technology, organizational change, and new employees and competencies are involved it can clearly be construed both as social and organizational innovation (Sundbo, 1998). In practical terms CRM implementation comprises the installation of advanced software to keep track of all customer-related information in a company. This makes the contact work more efficient and more integrated across different departments. Normally, information about existing customer, potential customers and suppliers are stored in one place. Information such as contact address, history of prior contacts, sales documentation, email correspondence and planned sales and contact activities become easily accessible. Enterprise Resource Planning systems (ERP) can be tapped as well as internet sources. Consequently this kind of company-wide software installation necessitates a complex re-organization of customer contacts within a company. Employees also need to be taught to handle and make use of the new software. Customers need to be informed about the information requirements and the mutual benefits of information sharing embedded in the CRM system. Hence, we argue that this kind of CRM implementation can be seen as an organizational innovation as new ideas, systems and work flows are introduced.

The theme for this publication is "Innovation with care." Hence, in this chapter we will single out the key idea of how customer care is managed. According to *Webster's Ninth New Collegiate Dictionary* (1990, p. 207) care can have several relevant meanings such as: painstaking and watchful attention, or being under someone's supervision. Also, care can take the meaning of having concern for someone or something. Our analysis will aim at illustrating the customer care concept in empirical settings.

Care will here be given three distinct meanings. First, care is understood as taking care, that is practically handling the interaction between buyer and seller for the benefit of all concerned. Second, care is also seen as caring, that is having the best interest of customers and employees at heart. This would imply that psychological sensitivity is used in organizing change and innovation. Third, care can also mean to be careful in the sense that appropriate safeguards and requirements in business are in place.

Hence we expect that rules and regulations are followed during innovation work and that systems are appropriately designed. In this chapter we will show how these three kinds of the care concept are applied during CRM innovation work. We have eclectically taken out illustrations from a case study on CRM implementation in Scandinavian firms. Three top executives responsible for the CRM operations were interviewed in early 2006 according to a semi-structured form. Interviews were taped and transcribed in English, which is the official working language for the group. Focus will first be on issues of CRM implementation and then on its effects on the performance of the organization.

IMPLEMENTATION OF CRM

How to select the right customer is the overarching problem when aiming your CRM innovation efforts. "It is about being out there" says the CEO of a small manufacturing company. "Define the key actors on the market and extrapolate who will be the winner. You want to stay with the winners." "It is very costly to get out of bad relationships and into new good ones" says marketing director of a large multinational industrial group. He continues: "In a developed market (with say 500 customers) it is straightforward understanding of your market. Make sure you know all the potential customers too, competitor's customers also." These excerpts show that the general market knowledge must be the starter for any CRM initiative or innovation.

How customers are targeted is a practical exercise. The CEO of the small manufacturing company says: "it all starts with screening your actual customers. Who has potential? We sit down with company management of key accounts and discuss their 3–5 year plans. We look at their strategic audit and talk about what they need. How can we help them?" This depicts several aspects of care. Taking care for the long term is essential. It makes co-ordination and planning more effective. By being open, caring becomes easier which is illustrated by the following citation. "Some big customers want to keep their distance, some are quite open about their plans." Caring becomes possible by instilling trust between partners. This can lead to more business as evidenced by the following expression from the marketing director of the multinational group: "we look at 100 per cent customer share. Sometimes we are two suppliers and one is going to have 75 per cent and the other 25 per cent." Hence customer care can lead to improved customer share with positive benefits for all.

A proactive approach in looking after the best interest of a customer is signaled by the following mode of operation by the CEO. "We always have a development portfolio of products (3–5) which we believe in. We introduce these to key account customers (3–4 years ahead). We do not ask them what they need (most of the time they do not know!). Instead, we show them our ideas and how they can be used for their operations. Then the customer will give back their own ideas about innovation." Hence by being the first to reveal the new ideas the CEO is taking care of the innovation process by initiating ideas and ways to co-operate. This will entice the customer to think along constructive lines of thought.

Another important CRM implementation issue is how one prioritizes customers. Industry respondents generally rank their customers according to several criteria such as: size, attractiveness of the market, market position and level of innovation. Then segmentation is carried out based on groupings from the rank order of customers. So called "crown jewels" are strategic accounts which must not be lost, no matter what. Key criteria are generally profitability and growth, with profitability the first priority. Often, loyal business partners share their growth plans. What is deemed crucial is the future potential of the customer. Implementing CRM, company resources must be rationally allocated to key customers. Normally, size and future potential define a key customer. Here a careful approach and evaluation drives internal resource allocation.

A basic tenet of CRM innovation is to minimize customer defections. By keeping customers the total income stream will be more controllable and costs to attract new customers less burdensome and crucial. However, once in a while, it becomes impossible to sustain a customer relationship. An example is given by the CEO of the small manufacturing company. "Recently, we lost a customer because of a shift in packaging type. We were not close enough to foresee this in time. What we normally do then is to push the termination of our relationship as far as we can into the future. Looking back our key account person was not close enough to read the situation and both his boss, the Commercial Manager, as well as myself should then have been more active." This evidence points to a lack of taking care of the long term prospect and also a lack of care in terms of having the customers' best interest at heart.

The contrary positive idea of care and care-taking is expressed by the marketing director of the small manufacturing company: "we need to know more about the customer operations than the customer himself!" He adds: "often the internal communication between internal departments is not so good. Our key account manager can act as a catalyst to improve this communication." This illustrates an inter-organizational process view of collaboration in which he sees his role as a go-between to enhance organizational communication and change.

Another key proposition of CRM implementation is to acquire new customers. Generally, more energy and time is needed to sign up a new customer than to maintain one. The role of key account managers is to evaluate the effort needed. Some account managers attempt to increase customer share, while other managers are also striving to attract new customers. Sometimes, incentive schemes are introduced to sales people to further these ends.

When practically implementing a new CRM system there are some fundamental requirements for success which are mentioned by all respondents. A first requirement is that top management takes the initiative and communicates the tangible value of CRM. That translates into looking out for and caring for the employees. When customers cry out for change, management must understand the urgent need of CRM implementation. Every system change needs an investment in time and effort, particularly from contact employees. Generally, they want to see a rather short-term payback of their efforts. Hence, employees should be given the tools to drive the CRM system themselves as much as possible. This is care-taking from the bottom-up of the system.

Often, barriers to successful implementation of CRM take the forms of a mismatch between the roles of marketing (frontline) and operations and production units. Often marketing functions cannot drive through change in the internal organization of the supplier and marketing people get caught in between. Customers may see little change and may complain to sales people who hear the same story over and over again. Hence, nobody takes care of the entire system. This is the prime reason for the attention and need of top leadership involvement.

The marketing director of the multinational industrial group sees certain phases that one has to move through when implementing CRM. "First, there is (1) *scepticism* later on (2) enthusiastic *revelation*. Then comes a period of (3) *confirmation*, the system starts to have effects, then *critique* (4) work is repetitive and we ask ourselves what are we getting out of it? Then (5) finally a *revitalization* makes us more focussed and dedicated to delivering customer value." The crucial thing according to him is "to make the front line understand and see the benefits of the CRM system. In the beginning it is going to be hard work to feed the system with information. Later on, frontline people will see that they gain much time in knowing more about the customer." This illustrates the need to take care of the entire work process in the organization to link up different parts and departments. This is done by caring for the concerns of the employees and being careful by taking a stepwise approach.

EFFECTS OF CRM ON PERFORMANCE OF THE ORGANIZATION

Generally, the focus in CRM operations is to increase the profitability of customers. Albeit, all customers may not be very profitable, seldom more than a few are unprofitable. These customers may be kept anyway, because of the difficulty in understanding the reasons behind the negative outcomes and what kind of corrective actions one could to take. This signals a careful approach in getting rid of unprofitable customers. Many customers will not remain profitable over time. Market difficulties or change of management may be the immediate causes. However, according to the marketing director of the multinational industrial group: "there is a higher probability that very profitable customers will remain profitable over time."

CRM also intends to enhance the long term loyalty of customers which is generally necessary for achieving a decent profitability in heavy industry. For example, it could take a minimum of one year, to two and half years to drive a customer's profitability from red to black numbers. Hence, by caring for customers in the CRM operations a loyalty effect is expected which in itself is a requirement for profit.

Increased business from each customer (customer share) is another expected effect of CRM. All respondents say they would like to have more of the business from existing customers. An example is given by the CEO of the small manufacturing firm.

> For one third of customers we may be having 100 percent of their business, in one third there may be a split between several suppliers, but with us having the main share. For the final third of customers we may have the small share of business. Therefore, it is essential to go in-depth with key people of the customer in order to be able to understand the drivers of development for tomorrow's needs. Key account managers have the prime responsibility to make certain we get the most out of our customers. Often our customers are quite open about the volume of business.

This citation illustrates the importance of being close to the customer in several ways. First, it is paramount to interact with and caring for key people, which means that social skills and effective interaction are important. Second, taking care is about looking ahead and jointly building an understanding about what drives the growth of the business by using the competencies of the customer. The CRM system needs to be interrelated with a human edge to be effectively updated.

Every now and then you lose a customer. This may not be a bad thing! Take the example of a customer going bankrupt. Then you lose future business and perhaps money. What respondents find important is to figure out, early on, for example in an emergent market, who will be number one on that market. It is essential not to lose such a one. Also, even if you lose a customer it is essential to understand why. Was it the decision of the customer, or something else that initiated the break-up? Let us not forget that sometimes it may be a good thing to lose a customer. For example, large customers which may yield close to zero profit, may stir up production and create extra costs and effort because of changes of order volumes. These examples illustrate that there are limits to the customer care aspects embedded in the CRM systems. Long-term estimations of winners and losers on a market are not within the scope of the system. Also, taking care of your own operations and outcomes may necessitate a termination of both loyal and long enduring customer relationships.

Measuring the impact of CRM is often very difficult. One such approach is discussed by the marketing director of the multinational industrial group.

> Every market and sales company makes an evaluation of the satisfaction of their truly loyal customers. Customers are asked to evaluate all frontline market staff such as key account managers and technical directors. It is part of a yearly in-house personal evaluation of the group. A number of questions pertain to: delivery, operations efficiency, service and marketing support. Bonuses and incentives are determined by the scores they get. All these scores are used as input at calculating the figures for the truly loyal customers.

Here, interestingly, the evaluations of the customers are also used to gauge their own loyalty apart from evaluating staff. The approach combines the care-taking of both customers and front line employees in one go! Numbers and evaluations are tracked over time to estimate the operations of CRM innovations and change.

All in all, CRM is expected to impact customer-oriented operations so that the company can deliver more value, that is a better price and a better offer because of accessing more information faster. This enhances the competitive advantage in general. The CEO gives a concrete example: "we may analyse ordering patterns and use that to suggest how they can better organize their purchasing." By customer care such as this example illustrates, the value for the customer is increased, which is the ultimate aim of CRM innovation and implementation.

CONCLUSIONS

We have selected a number of short illustrations from a case study on CRM implementation in Scandinavian firms. A few top CRM managers have been interviewed about their experiences with CRM. The focus has been on issues of CRM implementation and its effects on the performance of the organization. We have seen these efforts as a kind of organizational and social innovation. The focus of our interest has been the concept of customer care during innovation. We may conclude that three facets of customer care seem to be involved in CRM innovation work. Taking care is the physical handling of the CRM innovation process and running the subsequent CRM system. A number of examples have been given to illustrate the complexity of managing customer relationships. Caring for customers and employees is another, and equally important, facet of CRM innovation with care. This implies a psychological sensitivity to how customers and employees react when faced with change and re-organization. Clear aims and benefits from the CRM effort must be communicated from the top. Third, and finally, a careful operation means that safeguards are in place and care is taken when designing the system to ascertain that increased value can be offered to customers in exchange for the extra effort of data input and co-ordination. The issues of customer care analysed here will no doubt become even more topical when more and more companies begin to consider the recently developed notion of Global CRM.

REFERENCES

Bull, C. (2003), "Strategic issues in customer relationship management (CRM) implementation," *Business Process Management Journal*, **9** (5), 592–602.

Buttle, F. (2005), *Customer Relationship Management: Concepts and Tools*, Oxford: Elsevier Butterworth-Heinemann.

Chen, I. and K. Popovich (2003), "Understanding customer relationship management (CRM): people, process and technology," *Business Process Management Journal*, **9** (5), 672–88.

Colgate, M.R. and P.J. Danaher (2000), "Implementing a customer relationship strategy: the asymmetric impact of poor versus excellent execution," *Journal of the Academy of Marketing Science*, **28** (3), 375–87.

Doyle, P. (1995), "Marketing in the new millennium," *European Journal of Marketing*, **29** (13), 23–41.

Feinberg, R. and R. Kadam (2002), "E-CRM web service attributes as determinants of customer satisfaction with retail web sites," *International Journal of Service Industry Management*, **13** (5), 432–51.

Gummesson, E. (2004), "Return on relationships (ROR): the value of relationship marketing and CRM in business-to-business contexts," **19** (2), 136–48.

Haglund, L., B. Edvardsson and J. Mattsson (1995), "Analysis, planning, improvisation and control in the development of new services," *The International Journal of Service Industry Management*, **6** (2), 24–35.

Jain, D. and S. Singh (2002), "Customer lifetime value research in marketing: a review and future directions," *Journal of Interactive Marketing*, **16** (2), 34–46.

Johannisson, B. (1987), "Beyond process and structure: social exchange networks," *International Studies of Management and Organization*, 1, 3–23.

Kumar, V. and W. Reinartz (2006), *Customer Relationship Management: A Database based Approach*, New York: John Wiley & Sons.

Mattsson, J. and F. Orfila-Sintes (2007), "Innovation behavior in the hotel industry," *Omega*, in press.

Mattsson, J. and J. Rendtorff (2006), "E-marketing ethics: a theory of value priorities," *International Journal of Internet Marketing and Advertising*, **3** (1), 35–47.

Mattsson, J., C.F. Jensen, and J. Sundbo (2005), "An attractor-based innovation system in tourism: the crucial function of the scene-taker," *Industry and Innovation*, **12** (3), 357–81.

Mattsson, J., B. Ramaseshan and D. Carson (2006), "Let marketers reclaim corporate strategy," *Journal of Strategic Marketing*, **14** (June), 165–73.

Rigby, D. and D. Ledingham (2004), "CRM done right," *Harvard Business Review*, November, 118–29.

Sheth, J.N. and A. Parvatiyar (1995), "The evolution of relationship marketing," *International Business Review*, **4** (4), 397–418.

Sundbo, J. (1998), *The Organization of Innovation in Services*, Copenhagen: Roskilde University Press.

Sundbo, J., J. Mattsson, B. Millett and R. Johnston (2001), "Innovation in service internationalisation: the crucial role of the frantrepreneur," *Regional Development and Entrepreneurship*, **13** (3), 247–67.

Zeng, Y.E., J.H. Wen and D.C. Yen (2003), "Customer relationship management (CRM) in business-to-business (B2B) e-commerce," *Information Management & Computer Security*, **11** (1), 39–44.

4. Innovation with care in health care: translation as an alternative metaphor of innovation and change

John Damm Scheuer

INTRODUCTION

An innovation may be defined as an idea, practice or object that is perceived as new by an individual or other unit of adoption (Rogers, 1983: 11). Innovation in the public sector[1] may be divided into different types of innovations: a new or improved service, a process innovation, an administrative innovation, system innovation, conceptual innovation and radical change of rationality (Koch and Hauknes, 2005). The focus of this chapter is the introduction of a process innovation in the form of a new idea and object aimed at organizing clinical pathways for patients in health care organizations. The idea/object is that of a "clinical pathway" which may be defined as "a standardized, prewritten, one- or two-page document showing the interventions of all disciplines along a time schedule. In effect, it is a grid, with time as one axis and staff actions as the other" (Zander, 1995: 9). It was developed by the nurse Karen Zander and her team of nurses at the New England Medical Center Hospitals in Boston (Zander, 1995) (Garbin, 1995) (Hofmann, 1993). According to Karen Zander the idea of clinical pathways soon spread internationally because they were not conceptual, visually appealing, easy to write and effective in reviewing and decreasing the length of stay of patients (Zander, 1995: 11).

Following Vrangbæk (1999) the concept of New Public Management is related to a number of different tools of governance which, based on economic and theoretical ideas about rationality and incentives, recommends the use of governance mechanisms from private firms and the use of economic and market based incentive structures in public organizations. In Danish healthcare, New Public Management has resulted in a focus on quality, operational efficiency and reduction of costs. The means to these ends have been the introduction of governance tools such as standards, descriptions of "best practices" and rationally planned clinical pathways as well as the development of measurement systems, measuring performance related to quality as well as productivity.

As a consequence of this development it may be asked what innovation with care might mean in a situation where New Public Management ideas encounter practice in Danish health care organizations.

When innovative ideas are introduced in health care organizations, health care professionals such as doctors, nurses, managers and consultants often ask how an idea might be "implemented." When doing so, they often write and talk about implementation in ways that are in accordance with how researchers related to public policy and public administration have theorized the concept. In a book by the Danish Society for Quality Development in The Health Care Sector in order to educate doctors in quality development, two doctors define the concept of implementation as follows: "Implementation means the realisation of theories and plans in practice and is a general problem related to introduction of national legislation as well as local changes in clinical practice" (Kristensen and Kjærgaard, 2001: 234). Doctors in the Cochrane Collaboration as well as researchers in public policy and public administration have tried to develop an "evidence based" and general theory about implementation. A recent analysis of the literature and research show however that they have not been able to do so in spite of the considerable amount of research time and effort which have been put into doing research in organizational change in these research communities (Scheuer, 2007). Scheuer (2007) concludes that the reductionist attempts made by these researchers to understand the complex nature of organizational change have turned out to be less successful than one could have hoped for. Even though according to these researchers some evidence exists, it has turned out to be difficult to reach clear conclusions about the nature of and the cause and effect relationships related to organizational change in these communities.

The introduction of New Public Management ideas, originally developed in private firms, has been accompanied by a renewed interest and faith in rationally planned change in public health care organizations. The introduction of governance tools such as "Total Quality Management," "LEAN production," standards and descriptions of "best practices" and ideal clinical pathways as well as the focus on measurement systems and "evidence based medicine" may all be related to the idea of rationally planned change. The idea may be traced back to Frederick Winslow Taylor (1916) who suggested that there was "one best way" to perform a given task, that these ways should be investigated using scientific methods and that the labor ought to be divided between centrally placed experts who rationally planned how to perform certain tasks and workers at the shop floor level performing those tasks. As shown by Vikkelsø and Vinge (2004) this approach may however be criticized for several reasons. One may object ontologically, since it may be suggested that the incalculability of humans as well as contextual differences between one place and another will make it impossible to identify "the one best way" to organize a given task. One may also object that the division of the planning and performing of tasks is potentially unethical because the content, challenges and requirements of knowledge and independent thinking related to jobs are replaced by a demand for obedience to externally defined standards and systems.

As a consequence of the problems related to theorizing organizational change as implementation and planned rational change one may ask what might constitute an alternative and more appropriate theory of innovation and change in public health care organizations?

The alternative metaphor offered here is the metaphor of translation. According to the metaphor of translation, the diffusion in time and space of any token (innovative ideas, orders, artifacts or practices and so on) is assumed to be brought about by people each of whom may act in different ways: they may let the token drop, modify it, deflect it, betray it, add to it, or appropriate it (Latour, 1986). The paper suggests that the metaphor of "translation" may be a more relevant and empirically justified metaphor than the metaphors of "implementation" and "planned rational change" when studying processes of innovation and change in public organizations.

Thus, the two research questions focused upon here are:

1. How may innovation and change in public healthcare organizations be understood and theorized as a process of translation?
2. How may innovation with care be conceptualized in public health care organizations encountering New Public Management ideas?

The idea and case focused upon is the planning of clinical pathways in a subgroup that planned clinical pathways in a psychiatric ward at Frederiksberg Hospital – a hospital in the Copenhagen Hospital Corporation.

APPROACH TO THE STUDY OF INNOVATION AND CHANGE

This analysis is based on a becoming-realist ontology. Becoming-realists suggest that language actively configures entities and events in the world in the very act of representing. In our use of language we do not just "write about" our objects/subjects of analysis, but bring these objects into existence through representational acts of writing (Chia, 1996: 37). Research then is understood as artistically crafted pieces of work which do not describe but rather contribute to the making of organizational reality.[2]

As a consequence it is suggested that science has produced a number of powerful metaphors about organizational change which have been turned into a kind of "folk theories" (Larsen, 2001) about organizational change in Danish health care organizations. A metaphor may be defined as a figure of speech in which a word or phrase literally denoting one kind of object or idea is used in place of another to suggest a likeness or analogy between them (Encyclopaedia Britannica online). As pointed out by Morgan (1980) metaphors proceed through assertions that subject A is, or is like B, the processes of comparison, substitution, and interaction between the images of A and B acting as generators of new meaning. One metaphor identified is the metaphor of "implementation" (Scheuer, 2007). The other is the metaphor of planned (techno-) rational change (Borum, 1995). The two metaphors are described briefly in the following sections in order to allow a discussion of the relevance of these metaphors in relation to the innovation process which is identified empirically. The descriptions draw on a thorough analysis of the literature and research related to each metaphor made by Borum (1995), Morgan (1986) and Scheuer (2007). This is in order to pin-point accurately some main characteristics of these metaphors.

The analysis is performed using a model of translation that focuses on the "encounter" of innovative ideas/objects and local practice. It is followed by a discussion of the empirical characteristics of the innovation process having been analysed. It is suggested that the process is better understood if theorized as a translation process than as an implementation or rationally planned process. The innovation process is then discussed in order to define how innovation with care might be conceptualized in public health care organizations encountering New Public Management ideas. This if using the planning of clinical pathways in a psychiatric ward as an empirical example and if theorizing the process of innovation as a translation process.

The Metaphor of Implementation

Implementation has traditionally been understood as a part of the political process and especially the practical process of carrying out policies.[3] Parsons (1995) describes the policy process as a number of phases consisting of:

1. problem identification and definition;
2. presentation and evaluation of alternatives;
3. decision/choosing a policy;
4. implementing the policy; and finally
5. evaluating the policy.

According to Scheuer (2007) policy analysts and public administration researchers focus on the possibilities and limitations related to implementing legitimate policy decisions in practice. The construction of the organizational relationships are hierarchical. At the top of the hierarchy you have a (democratically or legitimately elected) decision center which set up legitimate policy goals. In the middle you find regional or local organizations or middle managers. And at the bottom you find street-level bureaucrats who are in charge of delivering services to the general public. Overall policy goals are subdivided into means-ends chains on their way through hierarchy. The direction in which ideas move as well as the type of processes identified are top-down, bottom-up or combinations of these two kinds of processes. Effective change is seen as dependent on predictive knowledge about which variables – top-down as well as bottom-up – are most likely to affect policy-implementation. What is perceived as a problem is when policy outcomes are not in accordance with policy goals.

The Metaphor of Planned Rational Change

The metaphor of planned rational change (Borum, 1995; Morgan, 1986) was developed in order to develop private firms but has been imported in health care organizations as a consequence of the managerialization of the health care sector (Vikkelsø and Vinge, 2004). In the (techno-) rational planning perspective, the organization is perceived as a machine which uses a rationally designed production system to transform inputs to outputs. The focus is directed at improving the efficiency, effectiveness and productivity of the organization. Ends related to outputs should be obtained as economically as possible.

Problems are perceived as design or system problems and solutions to these problems are dealt with by "redesigning," repairing or changing parts which seem to be dysfunctional. Managers plan and design necessary changes assisted by experts and analysts basing their decisions on rational calculations of costs and benefits as well as expected outcomes. The success of changes is measured in relation to the degree to which they are followed by improvements in the organizations output or in input-output ratios.

ANALYTICAL PERSPECTIVE

Translation in Actor Network Theory

Actor-network theory is based on a constructionist ontology (Latour, 1986; Callon, 1986; Callon and Latour, 1981; Latour, 2005). It claims that phenomena in the world are co-produced by humans and non-humans that is, material objects or things which may be said to be "authors of consequences" just as humans. Power, institutions and knowledge as well as other phenomena are all effects of the work of heterogeneous actor-networks. The concept of translation is associated with a network building activity where actors transform tokens (orders, ideas, artifacts and so on) into something else by translating or associating them with heterogeneous elements that might be humans as well as non-humans (objects). The diffusion in time and space of tokens (ideas, orders, artifacts) is assumed to be brought about by people each of whom may act or translate it in different ways: they may let it drop, modify it, deflect it, betray it, add to it, or appropriate it. Whether ideas or orders become powerful depends on the number of humans and non-humans that become (or do not become) associated with and start "acting on behalf of" them during the translation process. Or put differently; it depends on whether an actor-network is established or not in relation to a token (an idea, an order and so on). As a consequence, social phenomena such as institutions and/or power are an effect of the work of actor-networks rather than a cause in actor-network theory. What has to be explained is the process of translation and the establishment and (translation-) work of an actor-network that make such effects (institutions, power, knowledge and so on) possible. It is important to notice that the word "actor" in the word actor-network is given a semiotic definition. Actors are defined as "actants" that is humans and/or objects/things that are identified as being the source of an action. As an idea – defined as a token and an actant – is translated it, as well as the translators, are changed. Their identities are changed.

A Critique of Actor-network Theory

Some researchers (Scheuer, 2007) have criticized the ontological posture of actor-network theory.

According to actor-network theory, time and space are actants as other actants that might influence the translation process. Whether it does is seen as an empirical question. The ontology that results from this point of view seems in accordance with a society where the majority of the translation processes going on are taking place in cyberspace or through other means of communication making it possible to ignore the situatedness of the body and objects in time and space as well as distances in time and space. Even though such societies do exist – for instance in cyberspace – they are however still quite rare. And most people simply do not live their lives there! At least not yet! As a consequence another translation model is offered that combines a constructionist ontology and concepts of actor-network theory with an institutionalist perspective on organizational change that considers humans and objects "situatedness" in time and space as constitutive for the translation process. The model relates innovation with care with the careful bricolage needed when humans and non-humans doing different types of work are associated in order to construct innovations in organizations rather than with the more managerialist process of enrollment of actors suggested by Michel Callon (1986).[4]

Translation as Association

An idea is defined as "images which become known in the form of pictures or sounds (words can be either one or another)" (Czarniawska-Joerges and Joerges, 1996). The model of association (Scheuer, 2003) focuses on how innovative ideas are translated by being associated with heterogeneous elements – that is humans as well as non-humans (objects/things) – as the idea encounters local practice in organizations. Translation is defined as one of many alternative ways in which human actors can associate an element such as an idea with other human and non-human elements. The longer the list of humans and non-humans enrolled by an idea in a local time-space context as well as across these contexts, the more "durable" and powerful the idea becomes because the more humans and non-humans act on its behalf. It is suggested that translation processes have to be analysed by deciding upon when the translation process in focus starts and ends and by dividing the selected time-period into "episodes." An episode is "a number of acts or events having a specifiable beginning and end, thus involving a particular sequence" (Giddens, 1984: 244). The humans and non-humans being associated with the idea during each episode as well as across episodes are then analysed. In order to guide the analysis a number of questions are asked.

Breaking down the change process into episodes

1. Which projects are focused upon and are objects for translation activities during this process?
2. Which actors are responsible for these translation activities?
3. When do the translation activities start and when do they stop?
4. How may the process be organized chronologically into episodes on the basis of these activities?

Analysing the content and development during episodes

5. Which elements (actants) are associated with the idea in each episode?
6. Which kinds of events result in these elements being associated with the idea?

7. What are the relations between the translation activities going on in each episode?

A project is defined as one or several actors' "purpose at hand" (Schütz, 1973: 9). The process is subdivided into episodes by focusing on when actors translation activities related to idea related projects start and stop. Any event including conflicts among as well as between humans and non-humans influencing the content and development of the idea during and between episodes are identified. On the basis of the analysis, it is decided who the translators were, which kinds of translations they made and what kinds of effects that were the result of the translation of the idea.

Variables and conditions affecting the content and direction of the translation process are considered empirical questions which will have to be answered on the basis of empirical analysis. Existing theories might be used heuristically to set up hypotheses about which variables and conditions that might influence the translation process. The effect of the translation of an idea is assumed to be the result of work made by the established local as well as global collective of humans and non-humans acting on behalf of a token as an idea.

Institutionalization is associated with a relatively stable association of heterogeneous elements to an idea in time and space. De-institutionalization in contrast is associated with a dissolution of such a relatively stable association of heterogeneous elements in time and space. Stable associations of heterogeneous elements in time and space are referred to as translation bundles. When ideas have been institutionalized they appear as translations bundles – that is bundles of humans and non-humans – that actors can mobilize routinely in order to produce and reproduce organizational practices. The methods used in order to generate and analyse data is described elsewhere.[5]

ANALYSIS

Introduction to the Case[6]

When the Copenhagen Hospital Corporation was formed in 1995, a hospital plan was made that introduced several new ideas related to New Public Management. Quality development was one of these ideas. And clinical pathways was one of the concepts that was introduced in the Corporation's quality development plan. As a consequence of the hospital plan several new centers were established at the hospitals in the Copenhagen Hospital Corporation specializing in the specialties of basic surgery and medicine.

Frederiksberg Hospital was one of the smaller hospitals in the corporation. As a consequence of the Hospital Plan an Elective Surgery Center and a Center of Medicine was established at the hospital. Departments from the hospital as well as other hospitals were merged in order to establish these centers. The Hospital Management included the hospital's psychiatric ward in the reorganizing process because parts of their services were also to be reorganized as a consequence of the Hospital Plan. According to the corporation's quality development plan all centres and departments were expected to start reorganizing their patients' pathways and to plan them rationally in order to use resources more effectively and in order to improve the quality of patients' pathways. As a consequence – at Frederiksberg Hospital – subgroups focusing on quality development and clinical pathways were formed under the project's groups planning the establishment of the Elective Surgery Center, The Center of Medicine as well as the reorganizing of the psychiatric ward. The idea and case focused upon in the analysis is the planning of clinical pathways in the subgroup that planned clinical pathways in the psychiatric ward at Frederiksberg Hospital.

Breaking Down the Change Process into Episodes

April 1997–January 1998
In the first episode, the activities in the sub-group planning clinical pathways in the psychiatric ward focused on the quality development of clinical pathways in a way that was in accordance with the way the concept is presented in the literature about clinical pathways as well as by the Copenhagen Hospital Corporation. It was the psychiatrist formally appointed chairperson of the group and the representative from the hospitals Planning and Development department in the group who were the carriers of this translation. During the episode, the activities resulted in a discussion of the division of labour between different professional groups involved in the clinical pathway. The sub-group discussed whether the preconditions for describing an ideal clinical pathway for the acute psychiatric patient were met and whether the translation should be based on an alternative translation proposal based on an alternative "qualitative" understanding of quality development of clinical pathways. A proposal presented by the psychologist in the subgroup.

January 1998–October 1998
The second episode started when the head of the psychiatric department was met with internal and external pressures for progress in the work of the sub-group. As a consequence the psychiatrist who was the head of department took over the role as chairperson for the sub-group. The activities then set in motion included the formulation of a plan, discussions and

efforts to decide what kinds of elements would contribute to an ideal division of labour, coordination and communication between professional groups involved in the clinical pathway of the acute psychiatric patient.

October 1998–December 2000

In the third episode, the activities in the sub-group were directed toward producing the final report of the sub-group. A writing group consisting of a physiotherapist and a psychiatrist was established with the representative from the Planning and Development Department as assistant and the head of department as a "shadow member." The activities in the episode included the production of highly complicated flowcharts describing the clinical pathway, the formulation of quality goals related to the flowcharts, a round where the preliminary report was commented on by other hospitals as well as The Copenhagen Hospital Corporations representatives and the writing of the final report to the project group as well as The Copenhagen Hospital Corporation. The activities also included a number of development activities that were started up and related to the development of the clinical pathway for acute psychiatric patients.

Analysing the Content and Development During Episodes

First episode: April 1997–January 1998

An action plan is proposed   The subgroup planning clinical pathways in the psychiatric ward started out searching for knowledge about what a clinical pathway was. The psychiatrist who was chair person as well as the representative from the hospital's Planning and Development Department suggested that the goal was to construct a clinical pathway that was "evidence based" and based on standard principles for quality development of clinical pathways. It was expected that the descriptions of clinical pathways would contribute to improving the quality of treatment and care of patients as well as the efficiency and resource utilization of the department.

The chairperson started out wanting to include patients, patients' families as well as professionals outside the hospital in the planning of clinical pathways who were also involved. They were however never included in this work. The sub-group now decided to describe the ideal clinical pathway for the acute psychiatric patient because approximately 90 percent of the department's patients were acute patients. As a consequence, it was decided not to describe clinical pathways for different patients with different diagnosis such as "schizophrenia" or "affective disorder" or "depression." Instead it was the description of an ideal standard clinical pathway for acute psychiatric patients that was focused upon.

Construction of objects and discussion of division of labour   The sub-group wrote that the idea with the description was to coordinate the professionals involved central tasks and peripheral tasks in order to assure that the wishes and needs of patients were met and taxpayers "got value" for their money. The description of clinical pathways now resulted in the construction of different types of objects. The pathway of the acute psychiatric patient was described on paper. The pathway was divided into phases, each phase was defined and which tasks should be performed during each phase as well as how tasks were connected across phases were described. The work resulted however in a discussion of what the division of labour between professional groups ought to be in order to be ideal for the acute psychiatric patients. It was discussed which tasks of the different professional groups were central tasks and which tasks were more peripheral tasks. It was also discussed which groups of professionals were main actors in the different phases of the pathway and which of them should be allowed to meet the patient.

Especially the categorization of tasks into main tasks and peripheral tasks resulted in problems. But also the work of describing the ideal pathway itself turned out to create difficulties. As a consequence of this translation work, it became legitimate to open up a discussion of the division of labour between the different professional groups. Several professional groups including psychiatrists, psychologists, nurses and occupational therapists thought that their services were main rather than peripheral tasks in different phases of the clinical pathway if the pathway was to be an ideal pathway. While psychiatrists and sometimes nurses usually met with the patients at an early stage, more groups including psychologists and occupational therapists suggested that they should also meet the patient at an early stage in order to make things ideal. The group members were, however, not able to reach any final decisions about the ideal division of labour between the different professional groups in an ideal pathway for acute psychiatric patients.

Problems and conflicts over organizing principles   A psychologist in the sub-group now pointed out that the description of an ideal clinical pathway for acute patients would necessarily have to be based on common standards and principles for treatment of the patients in the department. The problem was, however, that such common standards and principles did not exist. Psychiatric patients would be treated applying a biological, psychological or social network approach to the treatment. Whether patients were treated with "pills" (biologically) or psychologically through dialogue depended however on where beds were available at the time when the patient entered the psychiatric ward. If the head of department where the patient got a bed was a biologist, the patient would be treated with pills. If he or she focused on psychological treatment and dialogue, the patient would be treated psychologically through dialogue. The psychologist pointed out that it would be impossible to describe, reorganize and develop the quality of clinical pathways in the psychiatric ward if some common standards and principles for treatment of patients were not formulated and agreed upon. Psychiatrists would have to accept treating patients biologically, psychologically or through social network therapy based on a careful examination of the patients specific needs instead of accidentally. As a consequence, the work in the subgroup was brought to a halt. The chair person of the subgroup told the head of the psychiatric ward and the Planning and Development Department that the subgroup would have to wait for the formulation of such common principles.

Progress is demanded and alternative perspective on clinical pathways is introduced At the end of the episode, the workgroup had created a problem that would have to be solved. It had to find a way of closing the discussion of the division of labor and of how to coordinate the different professional groups in an ideal way. The head nurse in the department and the Planning and Development Department now proclaimed that they were not satisfied with the progress of the work in the sub-group. According to them, the work had taken too much time and had produced too few results in relation to reorganizing clinical pathways in the psychiatric ward. At the same time, the psychologist in the subgroup produced a paper where she constructed two different perspectives on quality development of clinical pathways. One perspective was labeled "the quantitative approach" to quality development and was associated with the way clinical pathways were described in the literature about quality development. Where the "quantitative" approach assured the continuity and quality of pathways through measurements of how a ward performed in relation to certain predefined standards or goals the "qualitative" approach was suggested to assure continuity as well as quality through cross-disciplinary knowledge about treatment of psychiatric patients among involved professional groups. The psychologist suggested that the qualitative approach would be a more appropriate approach to quality development of clinical pathways of psychiatric patients and suggested that psycho-dynamic theory represented one possibility for establishing a common frame of reference.

Second episode: January 1998–October 1998

New chairman and a new plan As a consequence of the pressure from the head nurse in the ward and from the Planning and Development Department, the psychiatrist who was head of department took over the post as chairman for the sub-group. He made a new plan that was less ambitious. Instead of constructing an ideal description of the clinical pathway for acute psychiatric patients based on evidence based as well on local knowledge, it was decided to base it on local knowledge only. A conflict now arose between the chairman and the psychologist in the sub-group. The psychologist insisted on the importance of deciding according to which principles the clinical pathway of acute patients should be reorganized. The new chairman both listened and did not listen to her advice.

Improving information and processes of communication The chairman pointed towards a new direction for the work in the subgroup in his new and alternative plan. The group should decide which professional groups ought to meet the patients and who was supposed to do what when. Quality standards and goals should be formulated. And each professional group should describe the kinds of information it needed in order to be able to decide where the acute psychiatric patient ought to be placed for further treatment and to decide whether the professionals' own professional group ought to be involved directly in the treatment of a patient. The professional groups were also asked to describe the types of treatments that they could offer patients.

It was decided that the psychiatrist and nurse or other professionals who first met and collected information about the patient should collect information that was in accordance with the information needs thus described. This information would then be brought up and used as a basis for discussions and decision making at two kinds of meetings. A cross-disciplinary meeting where it was decided in what department the patient should be placed. And later a cross-disciplinary meeting in the department where the patient was placed aimed at deciding how to treat the patient. The chairman later explained that his translation proposal was based on the assumption that psychiatric patients are complex entities whose complexity is actually simplified by the biological, the psychological and the social-network perspectives on psychiatric illness. He suggested therefore that a cross-disciplinary perspective on the patient was the best way to guarantee the quality of the treatment of acute psychiatric patients. As a consequence it was the procedures related to collecting and processing information about patients that was focused upon. The consequence of the chairman's plan and translation proposal was that the discussion of the division of labour between professional groups ended.

Conflict about the role of nurses One conflict over the division of labour remained however. The chairman suggested that nurses as well as other professional groups should be able to assist the psychiatrist when he or she met with the acute psychiatric patient. This however was in conflict with the traditional division of labor where nurses used to assist the psychiatrists. The idea was that nurses as well as other professional groups should rotate in the function as assistants. The nurses however rejected the proposal and argued that the system suggested by the chairman was impossible since physiotherapists and occupational therapists were not legally allowed to delegate tasks to nurses and other professionals while nurses on their part were allowed to do so in all areas except in relation to tasks related to doctors/psychiatrists.

Third episode: October 1998–December 2000

Establishment of writing group and reintroduction of quantitative approach
In the third episode, the activities in the sub-group were directed toward producing the final report of the sub-group. A writing group consisting of a physiotherapist and a psychiatrist was established with the representative from the Planning and Development Department as assistant and the head of department as a "shadow member."

During the process of translation two different translations of quality development of clinical pathways had been introduced in the working group. One translation proposal was labeled "the quantitative" approach to quality development. The other was labelled "the qualitative" approach to quality development. The characteristics of these two translation proposals are summarized in Table 4.1.

As long as the sub-group focused on how the idea of clinical pathways could be translated internally in the ward, they were able to concentrate on internal discussions and activities aimed at improving coordination, communication and the division of labor between the different professional groups involved. They were also able to develop an alternative "qualitative" approach to development of clinical pathways. The writing of a report to hospital management and the representatives of the Copenhagen Hospital Corporation who had financed the project with 250 000 Danish Kroner, about €32 980, resulted in a change of this focus. The members of the subgroup as well as the writing group now became more aware of the external group's expectations of the group as well as to the outcomes of the group's work. The external group's were interpreted as being in favour of a quantitative rather than a qualitative approach to quality development.

The representative from the Planning and Development Department who had originally proposed the "quantitative" approach now reintroduced the perspective by making a proposal describing what the report to the external groups ought to contain. He suggested that it should contain

Table 4.1　　*Quantitative and qualitative approach to quality development*

Translation proposal	Quantitative approach	Qualitative approach
Translators	First chairman and representative from Planning and Development Department.	The psychologist and second chairman (Head of Department).
Problem to be solved	Creation of insight into the quality of clinical pathways in order to improve quality, effectiveness and the utilization of resources.	Creation of insight into information needs of employees in order to improve continuity as well as quality of clinical pathway.
Method	Construction of an evidence-based ideal clinical pathway, quality standards and goals, measuring of standards in relation to goals in order to analyse and improve clinical pathway if necessary.	Quality development of cross-disciplinary information used in decision making and of the processes and objects involved in co-ordination and communication between professionals.
Associated humans	Inside ward: Psychiatrists, psychologists, nurses, assistant nurses, physiotherapists, occupational therapists, welfare officers and secretaries. Outside ward: Representatives from local councils, general practitioners, patients and their relatives.	Inside ward: Psychiatrists, psychologists, nurses, assistant nurses, physiotherapists, occupational therapists, welfare officers and secretaries.
Associated non-humans	Papers in A4 format defining and describing phases as well as tasks in each phase. Flowcharts and quality standards/goals on paper.	Papers in A4 format defining and describing phases as well as tasks in each phase. Reports about information needs and treatments of involved groups of professionals.

a description of the purpose of the clinical pathway, a schematic description of the clinical pathway for acute psychiatric patients, a description of goals and standards related to phases in the pathway, an explanation of how goals and standards were to be measured and documented as well as an explanation of how these activities were to be organized. As a consequence, the activities of the writing group as well as the subgroup were now directed at producing these elements to the report.

A purpose of the clinical pathway was described and a huge and complicated flowchart describing the ideal pathway for acute psychiatric patients was produced. Goals and standards were described for each phase as well as for relations between phases. When the representative from the Planning and Development Department pointed out that the standards and quality goals were not in accordance with the literature about quality development in health care, they were changed so that they were. The different professional groups as well as the group of middle managers – that is psychiatrists – however made it clear that the goals and standards were not binding but only suggested as advice about how such goals might look like. As a consequence the flowcharts and goals that had been set up in relation to the pathway were turned into symbols of "the way to do things" rather than objects that had a binding effect on clinical practice.

Report commented by external groups and development activities are started up   In December 1999, the group's report was sent to the Copenhagen Hospital Corporation, other hospitals and psychiatric wards for comments. The report was only changed marginally as a consequence of these comments. The report could be read as if the subgroup had lived up to standard expectations and had applied a quantitative approach to the development of clinical pathways for acute psychiatric patients. When scrutinized in detail however it turned out that the report was not intended to bind clinical practice in the way that the report seemed to suggest. Flowcharts had been made and standards and goals had been set up in relation to each phase of the pathway. But they were "examples" and not necessarily supposed to be put into practice in the ward. In the report the purpose of the clinical pathway was described in a way that was in accordance with the qualitative approach to development of clinical pathways. Moreover the activities that were supposed to be started up as a consequence of the report did not include setting up a measurement and info "feed-back" system that would make it possible to measure and develop pathways on the basis of measurements related to standards/goals.

The activities that were supposed to be started up according to the report were rather professional and development activities that were related to the qualitative approach. The report pointed out that a common frame of reference for treating psychiatric patients should be developed, that the cross-disciplinary culture should be strengthened and that the different professional groups ought to be supervized by psychologists and educated. It also pointed out that the structure and content of meetings should be improved and that the necessary information and tasks should be collected and/or performed. The department should moreover attempt to live up to time limits and assure that handover procedures as well as tasks were performed in accordance with expectations. The processes however should not be monitored using a measuring system.

Other more peripheral activities were started up and related to the report and the work done in relation to clinical pathways. The activities included the establishment of a group that should develop a concept for treatment plans, a group about psycho-education of patients and their relatives. A group that should suggest a plan for meetings in the ward that were not related to the treatment of patients. The information given to patients entering the ward should also be revised as should administrative procedures related to entering data into data systems in the ward.

Effects of the Translation Process

In the first episode, the subgroups division of labor into central and peripheral tasks as well as the construction of a written and "ideal" description of the clinical pathway for acute psychiatric patients opens a discussion of the ideal division of labor between professionals involved in the pathway. The effect is that the division of labor which have until then been black-boxed (Latour, 1987) and represented tacit knowledge is turned into an object of discussion. The discussions result in a time delay. The psychologist enters the discussion and suggests that the quantitative approach to quality development of clinical pathways is insufficient in relation to psychiatric patients. She proposes a qualitative approach as an alternative. She also points out that the psychiatrists in the ward do not agree on how to treat the patients making the construction of standardized pathways difficult. As a consequence or an effect the chairman informs the project group and the Planning and Development Department that the group need more time to finish its work.

In the second episode, the role as chairman is taken over by the psychiatrist who is head of department because the head nurse and the Planning and Development Department is dissatisfied with the lack of progress in the sub-group's work. The chairman makes a written plan and suggests that quality development of pathways has to be based on a cross-disciplinary perspective on psychiatric illness since the psychiatric patient is more complex than suggested by the psychiatrist's biological, psychological and social-network perspectives on psychiatric illness. The chairman constructs his own version of the qualitative approach to development of pathways instead. What is focused upon is quality development of cross-disciplinary information used in decision making and development of the processes and objects involved in cross-disciplinary coordination and communication between professionals. The objects produced during the episode include descriptions/reports about the information needs of involved professional groups as well as descriptions of the types of treatments each group can offer the patients. The activities also include a specification of the division of labor between one or two professionals who meet the acute psychiatric patient and two cross-disciplinary teams which uses the information collected by these persons to decide where to place and how to treat the patient in focus. Since the cross-disciplinary perspective makes all professional groups equally important, the effect of this translation proposal is that the discussion of the division of labor between professional groups and the division of tasks into main and peripheral tasks are closed down. Another effect is that an alternative perspective on quality development of pathways is developed which is adjusted to local knowledge as well as to professional interests.

In the third episode the representative from the Planning and Development Department makes the writing group and the head of department aware that external groups might have a quantitative rather than a qualitative perspective on quality development of pathways. As a consequence the activities in the subgroup are directed at producing flowcharts and standards/goals related to the pathway of acute psychiatric patients. The effects of these flowcharts and goals are however limited since the professional groups and managing psychiatrists make sure that standards and goals are only mentioned as "examples" in the final report and not as something that is necessarily supposed to be put into practice in the ward. As a consequence the subgroup's final report may be described as a hybrid. On the one hand it may be read as in accordance with the quantitative approach to quality development of pathways since it includes elements such as flowcharts and standards/goals normally related to this approach. On the other hand, the report sums up the work done in the subgroup on the basis of the alternative "qualitative" approach. The development activities which follow up on the report are however related to the qualitative rather than the quantitative approach. As a consequence, the effect of the translation process is that a report is produced that legitimizes the activities that have been going on in the ward in relation to external actors as well as sum up the results of the development activities internally. The report also has effects since it serves as a point of departure for and a way to legitimize new development activities which are related to the qualitative rather than quantitative approach to quality development of pathways.

DISCUSSION

Innovation as an Implementation, Rationally Planned or Translation Process?

As the analysis has demonstrated both humans and non-humans are mobilized and associated by the translators in order to construct a local translation proposal that translates an "incoming" idea – the idea of clinical pathways – in a way that is in accordance with local knowledge, organizing processes as well as professional interests. Thus as a consequence of the "movement" of the idea in time and space and the local translation work and learning process which is an inevitable part of that movement the idea as well as the translators of the idea are changed.

The metaphor of implementation however is based on the assumption that legitimately decided policy ideas may move relatively unchanged from a decision center through hierarchy to the street level bureaucrats who will then act "on behalf" of the idea. Top down as well as bottom up elements influencing the movement of ideas may be identified and to a certain degree controlled making this movement of a relatively unchanged idea possible. When conceptualized as translation however, the movement of policy as well as other ideas on the contrary may be interpreted as a movement of an actant in time-space from one community of translators doing translation work to another. As demonstrated in the case study, the idea may be seen as moving not by its own force or by itself as implied in the metaphor of implementation but as a consequence of humans that choose to "pick up" and start acting on behalf of the idea by trying to mobilize and associate humans and non-humans to the idea.

The metaphor of planned rational change may be questioned on the basis of the analysis as well. The metaphor suggests that organizations may be understood as "machines." Problems are dealt with by "planners" placed in decision centers relatively far away from the organizing processes that they attempt to change. Problems are perceived as design or system problems which are dealt with by "redesigning," repairing or changing parts which seem to be dysfunctional. According to the analysis however it may be suggested that it is the translation work of local translators that turn out to be important for the change process rather than remote planners work. It may also be suggested that ideas and concepts do not move easily in time and space. Decisions of experts, administrators and politicians about the introduction of certain translations of ideas or concepts such as "clinical pathways" may have effects by directing local actors' energy and activities in certain directions. But what the study shows is that whether an idea or elements of an idea are introduced in practice depends on whether someone chooses to translate it – that is to mobilize and associate humans as well as non-humans to the idea in a way that makes them act on behalf of that idea. As demonstrated in the analysis, it was the in-depth local knowledge of the head of department and the psychologist that eventually made development and change possible in the form of an invention – the qualitative approach to development of pathways. As a consequence it seems likely that planners placed remotely from local organizing processes will be less able to obtain the level of detailed insight needed to translate ideas and change local organizing processes than translators who are near in space and time.

The metaphor of planned rational change may be criticized for implying that there exists one best way of organizing a particular activity. The study indicates however that the unpredictability of humans and non-humans as well as the complexity of translation processes makes it unlikely that there exists "one best way" of organizing an activity. There may rather be more or less successful ways of translating an idea given the humans and non-humans that happens to be at hand. Instead of planning rationally and top-down in order to achieve goals, it may therefore be as appropriate to initiate learning processes where humans and non-humans are associated or translated heuristically until "wished for" and attractive effects occur. If you want to change well established organizing practices in your department or organization, you may get ideas about how to do so by skipping standard ways of thinking about what is going on in the organization. By just observing the kind of work done by humans and non-humans making certain facts or phenomena possible in the organization (Vickelsøe and Vinge, 2004). It may be possible to come up with more locally grounded and intelligent ideas about how to change the organization in "wished for" directions than by using the standard and decontextualized ideas being promoted under the label of "New Public Management."

It may now be concluded that the metaphors of implementation and rationally planned change turn out to be inadequate and unrealistic when processes of organizational innovation and change are scrutinized in detail. Even though the metaphors might be helpful for politicians, public administrators and professionals in attaching meaning to the political, decision and planning processes in which they take part they are not helpful if one wants a better understanding of the complex and local processes where policy ideas encounter practice and produce or do not produce effects that may turn out to be intended as well as unintended (Giddens, 1984). If you want an in-depth understanding of the processes that were observed in the analysis they may be better understood if theorized as translation processes.

Innovation with Care when NPM Ideas meet Practice

The innovation process that have been analysed may now be discussed in order to define how innovation with care might be conceptualized in public health care organizations encountering New Public Management ideas. That is to say if using the planning of clinical pathways in the psychiatric ward as an example and if theorizing the process of innovation as a translation process.

The idea that clinical pathways were related to a description with time as one axis and staff actions as the other (Zander, 1995: 9) was maintained in both the quantitative and the qualitative approach. All other elements associated during the translation process analysed were however different making the idea of clinical pathways different. As demonstrated, the translators in the subgroup planning clinical pathways in the psychiatric ward rejected the quantitative approach to quality development of pathways since it was not in accordance with local knowledge, organizing processes as well as professional interests.

The clinical pathways of patients having broken a leg may be standardized and subjected to measurements related to quality goals since such patients are easy to diagnose and the appropriate treatment is well known. Acute psychiatric patients however were constructed as complex entities that were difficult to diagnose and treat in a standardized way. As a consequence, the psychiatrist and the psychologist suggested that a cross-disciplinary perspective on the patient was needed and that it was quality development of cross-disciplinary information used in decision making and of the processes and objects involved in coordination and communication between professionals that should be focused upon.

On the basis of the analysis, the concept of innovation with care may therefore be defined as local translators' translation of innovative ideas in a way that tests the pros and cons of an idea in relation to local knowledge and modes of organizing and take appropriate steps to integrate those elements of the idea that might improve local performance and treatment of patients while avoiding and buffering of those parts of an idea that are seen as harmful or dysfunctional on the basis of local knowledge and ways of organizing.

It may be suggested that the translators in the psychiatric ward showed care by protecting local organizing processes from harmful and dysfunctional disturbances that might have been caused by a concept – the quantitative approach – that was ill fitted to the kinds of patients, tasks and modes of organizing that characterized the psychiatric ward. They also showed care toward patients who were seen as treated better if applying a qualitative rather than quantitative approach to quality development of pathways.

Finally, the translators showed care by protecting the politicians and public administrators who "do not know what they do" from possible harmful effects of their decisions.

Keeping the last point in mind, it may be argued that the analysis contributes by helping to open up the "black box" of decoupling (Orton and Weick, 1990; Meyer and Rowan, 1991). On the basis of the study it may thus be suggested that there may be many good reasons that explains why policy ideas and other innovative ideas are seldom implemented as intended (Pollit, 2001). And that professionals' innovation with care may be one of those reasons.

Innovation with care may be defended ethically as well as legitimized by health care professionals by referring to medical ethics which places the interests of patients higher than rational and economic interests of politicians and administrators. On the basis of the analysis, it may be argued that the sets of techniques and programs found to improve mean scores of some groups such as measurement systems related to clinical pathways were not sensitive enough to the differences of skills and strengths among those professionals who served nor individual differences among those psychiatric patients served (Polkinghorne, 2004).

It should be noted that health care professionals attempts and efforts to innovate with care may sometimes be an inhibitor of change and be better understood as a defence of illegitimate professional interests. In the case, the translation of the idea of clinical pathways into an ideal pathway for acute psychiatric patients resulted in a long discussion of the division of labor between different professional groups. A theme that had until then been silenced or blackboxed (Latour, 1987) in the ward. The discussion and problem of the division of labor was solved by the second chairman when he introduced a translation proposal that made all professional groups equally important in the clinical pathway by emphasizing a cross-disciplinary perspective on the patient and the organizing process. As a consequence attention to and experiments with more industrialist ways of organizing the pathways that may have been advantageous for patients as well as employees were neglected.

CONCLUSION

The idea and case focused upon in the analysis was the planning of clinical pathways in a subgroup that planned clinical pathways in a psychiatric ward. The analysis was performed using a model of translation. Translation was defined as one of many alternative ways in which human actors can associate an element such as an idea with other human and non-human elements. The translation process was analysed by dividing the innovation process into episodes. The humans and non-humans that were associated with the idea during each episode as well as across episodes were then analysed.

The innovation process consisted of three episodes. In the first episode a discussion of the ideal division of labor related to acute psychiatric patients and the psychologist's dissatisfaction with the "quantitative approach" represented by the chairman and representative from the Planning and Development Department led to construction of an alternative "qualitative approach" to quality development of clinical pathways. It moreover led to a time delay. In the second episode, the time delay resulted in the head of department – a psychiatrist – taking over the role as chairman of the subgroup. He made a plan that was followed by construction of yet another translation proposal based on the "qualitative approach" (Table 4.1). As a consequence the conflict over the division of labor between professionals in the group was solved and an alternative perspective was developed that was adjusted to local knowledge, organizing processes as well as professional interests. Finally in the third episode the activities were directed at producing the subgroup's final report that was formulated as a hybrid between the quantitative and qualitative approach to quality development of clinical pathways. The report thereby legitimized the activities in the subgroup *vis-à-vis* external groups as well as summed up the results of the development activities in the subgroup internally. The report moreover pointed out and legitimized further development activities that were primarily related to the qualitative approach to development of pathways.

On the basis of the analysis it is concluded that the innovation process observed may be better understood if theorized as a translation rather than an implementation or rationally planned process.

It is furthermore concluded that the concept of innovation with care may be defined as local translators' translation of innovative – New Public Management or other – ideas in a way that tests the pros and cons of an idea in relation to local knowledge and modes of organizing and take appropriate steps to integrate those elements of the idea that might improve local performance and treatment of patients while avoiding and buffering of those parts of an idea that are seen as harmful or dysfunctional on the basis of local knowledge and ways of organizing.

In relation to the process of innovation the study implies more bottom-up and less top-down involvement of actors in innovation processes and an emphasis on creating local flexible processes of learning aiming at producing certain attractive effects by experimenting with heuristically combined humans and non-humans rather than aiming at implementing certain (New Public Management or other) ideas and concepts in certain ways. Instead of asking whether certain results related to means-ends relations proposed by ideas and concepts are met translators should ask whether the translation process results in positive or "wished for" rather than negative effects. This is because the effects of introducing an idea or concept is dependent on the mobilization of humans and non-humans that may act differently and may be combined differently from one time-space context to the other.

There are several practical consequences of this study. The study indicates that the unpredictability of humans and non-humans as well as the complexity of translation processes makes it unlikely that there exists "one best way" of organizing an activity. There may rather be more or less successful ways of translating an idea given the humans and non-humans that happens to be at hand. Thus, instead of planning rationally and implementing top-down in order to achieve goals it may therefore be appropriate to initiate learning processes where humans and non-humans are associated or translated heuristically until "wished for" and attractive effects occur. An innovative idea as clinical pathways may be used by top management to direct activities in an organization in a certain direction. Top management may also point out the types of effects or outcomes that it expects and construct systems making it possible to measure those outcomes. It may also point out certain ideas, objects or tools that it believes might help employees achieving those outcomes as long as the adoption of them is voluntary for employees who will know better which ideas, objects or tools might be helpful for them. Top management should moreover focus on supporting local learning processes rather than insist on introducing standardized (New Public Management or other) ideas in standardized ways.

NOTES

1. The public sector includes all organizations in the field of the public administration, social security, law and order, education, health care, and social and cultural services, irrespective of their funding source and the legal form of the supplier (Koch and Hauknes, 2005: 17).
2. Being-realism is a fundamental ontological posture which asserts that reality pre-exists independently of observation and as static, discrete and identifiable "things," "entities," "events," "generative mechanisms" and so on (Chia, 1996). In contrast becoming-realism gives primacy to a processual view of reality. How an "entity" "becomes" constitutes what that actual entity is so that the two descriptions of an entity are not independent. Its "being" is constituted by its "becoming" (Chia, 1996: 33). Becoming-realists suggest that language actively configures entities and events in the world in the very act of representing. In our use of language we do not just "write about" our objects/subjects of analysis, but bring these objects into existence through representational acts of writing (Chia, 1996: 37). Research may be understood then as artistically crafted pieces of work which do not describe but rather contribute to the making of organizational reality (Chia, 1996: 45–6). As a consequence different researchers claim that they have a privileged understanding of

social events and phenomena in the world is substituted by a situation where such claims become just one "voice," just one type of (constructed) knowledge claim among others taking part in the making of organizational reality.

3. According to Heclo "policy" is "a course of action intended to accomplish some end" (Heclo, 1972: 84). Politics deal with the allocation of resources and values in society or as Harold Lasswell puts it: "Politics; Who Gets What, When, How" (Lasswell, 1958).

4. Michel Callon (1986) describes the process of enrollment as consisting of four phases: 1. The phase of problematization where a translator defines a problem that may interest and mobilize human as well as non-human actors, 2. Association of humans and non-humans by using devices of "interessement," 3. A negotiation with human and non-human actors concerning the role that they are to play in the innovation, 4. Mobilization of associated humans and non-humans through their representatives.

5. Since employed as a development consultant at Frederiksberg Hospital at the time I took part in almost all meetings in the subgroup that planned how to introduce clinical pathways in the hospital's psychiatric ward. In order to deal with this I started out writing my own "story" about the work in the subgroup. This in order to turn this story into data that could be studied as other types of data about the process in the subgroup. The three and a half year long process was carefully documented through a huge amount of documents such as elaborate resumées of discussions going on at meetings as well as documents produced and presented at meetings. As a former member of the group and as the secretary of the group I had access to as well as knowledge about all relevant documents and persons involved in the development and innovation process in the group. These as well as related reports and archieval materials were collected. Once collected the materials were organized by writing significant events related to the translation of the concept of clinical pathways into a chronologically organized document that gave as complete an overview over the change process as the documents collected allowed. The process described was then broken down into episodes and the content and development during each episodes was analysed asking the seven questions mentioned in the translation model above. Since the process in focus was well documented in written documents and well known since I had taken part in the process personally no interviews were performed. Such interviews could however have further strengthened the empirical basis of the analysis since my recollection of the process as well as my interpretation of the written documents related to the process may have been biased. The resumées of the meetings held in the working group were however read and accepted by all members of the group at the following meeting which strengthen the credibility of these documents.

6. The introduction to the case is based on Scheuer (2003).

REFERENCES

Borum, F. (1995), *Strategier for organisationsændring*, Copenhagen: Handelshøjskolens Forlag.

Callon, M. (1986), "Some elements of a sociology of translation: domestication of the scallops and the fishermen of St. Brieuc Bay," in J. Law (eds), *Power, Action and Belief – A New Sociology of Knowledge?*, London: Routledge.

Callon, M. and B. Latour (1981), "Unscrewing the big Leviathan: how actors macro-structure reality and how sociologists help them to do so," in K. Knorr-Cetina and A.V. Cicourel (eds), *Advances in Social Theory and Methodology: Toward an Integration of Micro and Macro-sociologies*, Boston, MA: Routledge & Kegan Paul.

Chia, R. (1996), "The problem of reflexivity in organizational research: towards a postmodern science of organization," *Organization*, 3 (1), 31–59.

Czarniawska, B., and B. Joerges (1996), "Travels of ideas," in B. Czarniawska and G. Sevón (eds), *Translating Organizational Change*, Berlin and New York: Walter de Gruyter.

Czarniawska, B. and G. Sevón (1996), *Translating Organizational Change*, Berlin and New York: Walter de Gruyter.

Garbin, B. (1995), "Introduction to Clinical Pathways," *Journal for Healthcare Quality*, **17** (6), 6–9.

Giddens, A. (1984), *The Constitution of Society – Outline of the Theory of Structuration*, Cambridge: Polity Press.

Heclo, Hugh (1972), review article: "policy analysis," *British Journal of Political Science*, **2** (1), 83–108.

Hofmann, P.A. (1993), "Critical path method: an important tool for coordinating clinical care," *Journal of Quality Improvement*, **19** (7), 235–46.

Kristensen, F.B. and J. Kjærgaard (2001), "Implementering og kvalitetsudvikling," in J. Kjærgaard, J. Mainz, T. Jørgensen and I. Willaing, *Kvalitetsudvikling i sundhedsvæsenet*, Copenhagen: Munksgaard.

Koch, P. and J. Hauknes (2005), *On Innovation in the Public Sector*, Oslo: Publin Report No. D20, NIFU STEP.

Larsen, B. (2001), "Implementering – det værste vrøvleord i mands minde," in S. Jönsson and B. Larsen (eds), *Teori & Praksis. Skandinaviske perspektiver på ledelse og økonomistyring*, Copenhagen: Jurist- og Økonomforbundets Forlag, pp. 361–405.

Lasswell, Harold D. (1958), *Politics: Who Gets What, When, How*, Cleveland, OH: Meridan Books.

Latour, B. (1986), "The powers of association," in J. Law, *Power, Action and Belief – a New Sociology of Knowledge?*, London: Routledge & Kegan Paul.

Latour, B. (1987), *Science In Action*, Cambridge, MA: Harvard University Press.

Latour, B. (2005), *Reassembling the Social – an Introduction to Actor-Network-Theory*, Oxford: Oxford University Press.

Meyer, J.W. and B. Rowan (1991), "Institutionalized organizations: formal structures as myth and ceremony," in W. Powell, P.J. Dimaggio (eds), *The New Institutionalism in Organizational Analysis*, Chicago, IL and London: The University of Chicago Press, pp. 41–62.

Morgan, Gareth (1980), "Paradigms, metaphors, and puzzle solving in organization theory," *Administrative Science Quarterly*, **25** (4) (December), pp. 605–22.

Morgan, G. (1986), *Images of Organizations*, London: Sage.

Orton, J.D. and K.E. Weick (1990), "Loosely coupled systems, a reconceptualization," *Academy of Management Review*, **15** (2), 203–23.

Parsons, W. (1995), *Public Policy: An Introduction to the Theory and Practice of Policy Analysis*, Aldershot, UK and Brookfield, VT, US: Edward Elgar.

Polkinghorne, D.E. (2004), *Practice and the Human Sciences – The Case for a Judgment-Based Practice of Care*, Albany, NY: State University of New York Press.

Pollit, Chr. (2001), "Convergence: the useful myth?", *Public Administration*, **79** (4), 933–47.

Rogers, E.M. (1983), *Diffusion of Innovations*, New York: The Free Press.

Scheuer, J.D. (2003), "Patientforløb i praksis – en analyse af en idés oversættelse i mødet med praksis", Handelshøjskolen i København, Januar 2003 (phd afhandling).

Scheuer, J.D. (2007), "Intervention, implementation and translation as metaphors of change in health care and public organizations – what researchers know and think they need to know," in S. Scheuer and J.D. Scheuer (eds), *The Anatomy of Change*, Copenhagen: CBS Press (forthcoming).

Schütz, A. (1953/1973), "Common-sense and the scientific interpretation of human action," in A. Schütz, *Collected Papers*, vol. 1, The Hague: Martinus Nijhoff, pp. 3–47.

Taylor, F.W. (1916), "The principles of scientific management," in J.M. Shafritz and J.S. Ott (1996), *Classics of Organization Theory*, New York: Harcourt Brace College Publishers.

Vikkelsø, S. and S. Vinge (2004), "Optimeringsparadigmet og dets begrænsning," in S. Vikkelsø, and S. Vinge (eds), *Hverdagens arbejde og organisering i sundhedsvæsenet*, Copenhagen: Handelshøjskolens Forlag.

Vrangbæk, K. (1999), "New Public Management i sygehusfeltet – udformning og konsekvenser," in E.Z. Bentsen, F. Borum, G. Erlingsdóttir and Kerstin Sahlin-Andersson (eds), *Når styringsambitioner møder praksis – den svære omstilling af sygehus- og sundhedsvæsenet i Danmark og Sverige*, Copenhagen: Handelshøjskolens forlag.

Zander, K. (1995), *Collaborative Care: Two Effective Strategies for Positive Outcomes*, Chicago, IL: American Hospital Publishing Inc.

PART 2

Importance

5. The public library between social engineering and innovation with care

Lars Fuglsang

INTRODUCTION

How can innovation be organized and planned in the public sector? This chapter discusses four frameworks (or modes) of innovation each of which has consequences for the organization of innovation. The four frameworks are: entrepreneurial innovation, institutional innovation, open innovation and strategic reflexive innovation. After introducing these, the chapter goes on to investigate the framework for innovation in a particular domain of innovation, namely the Danish public library system. In this case it is argued that the framework for innovation is changing from the institutional mode to open innovation.

The chapter argues that open innovation requires a careful and imaginative balancing of two forces of innovation: variation and selection (see Nelson and Winter, 1977). This implies that variation and selection become the important mechanisms of organizational change rather than, for example, homogenization (as implied by DiMaggio and Powell, 1983). First, open innovation creates a need to take care of and incorporate many different ideas that work in a specific domain such as the library. Second, open innovation requires a strategic field where some opinions and ideas are selected and scaled up as being, in some senses, more relevant, effective or important than others. "Care" means the attempt to maneuver carefully and imaginatively between these two sides of innovation that otherwise can pull in different directions. The chapter argues that one relevant way in which selection can be carefully combined with variation in the case of today's public library is in the shape of a "social and strategic arena."

The approach to innovation which is adopted in this chapter is, therefore, a systemic approach which has three components, the domain of innovation (the single library), the framework (or mode) of innovation, and the field of innovation (the broader selective environment of the library) (for the concepts of the domain and the field see also Csikszentmihalyi, 1996). Innovation with care implies that attention is paid to variation and selection rather than to homogenization among single public institutions or organizations.

INNOVATION AND THE PUBLIC LIBRARY

In innovation theory, innovation is understood as an economic and social activity of which the purpose is to add value (economic or social) to stakeholders as well as society as a whole (economic growth). For those who want to innovate, the question becomes one of exploring where and how this value can be generated by developing goods, services or, in the more recent literature, experiences (for various definitions of innovation, see the introduction to this book).

According to this approach, innovation is not necessarily something that is easily planned in a laboratory and spreads from there to the market. Increasingly, innovation is seen as an interactive process that involves many and changing actors in various forms of heterogeneous networks and partnerships. The main reason for this is that innovative resources have become much more widely distributed throughout society than just a few decades ago. Consumers have become more critical and demanding, and business partners, competitors, consultants and users more important to the company's innovation process. Innovation has therefore become a more common activity that involves many people at many levels. This means that the organization of innovation must be much more "open" and "heterogeneous" today than just a few decades ago, where the innovative resources were concentrated in the companies' own research and development laboratories or in public institutions.

The public library is a domain of innovation in the public sector that provides an interesting example of open innovation. I will argue that the Danish public library is not simply a bureaucratic public institution, which is conservative and reluctant to change, but innovative and entrepreneurial, able to engage employees and others in communication about innovation. In addition to this, I argue that established approaches to change in public organizations, such as institutional theories stressing isomorphism, homogenization and decoupling effects (DiMaggio and Powell, 1983), simply seem insufficient as a framework for understanding the change processes that penetrate the public sector. It is simply not true, as claimed by some of these theories, that there is lack of innovation with little variation and selection in the public sector in favor of homogenization processes. Rather, in many places, innovation, variation and selection are becoming important mechanisms of change (see also Fuglsang, Højland and Storm Pedersen, this volume).

The example of the public library shows both the growing impact of open innovation and how contemporary organizations are trying systematically and proactively to involve employers and users in the innovation process in order to respond to the requirements of open innovation.

INNOVATION WITH CARE

Innovation with care is an approach to innovation which points to the need for careful and imaginative balancing of variation and selection: it emphasizes the need to involve many opinions and ideas in innovation, that is, opinions about "ideas that work" in a given domain (Young Foundation, 2006). And it stresses the wider diffusion and selection of some of these ideas in an imaginative rather than mechanical way. This selection and diffusion is understood as based not only on the market mechanism but also on other mechanisms of social interaction and diffusion (see Rogers, 1995). Selection and variation are understood as mechanisms of organizational change that are different from homogenization.

Often, for public institutions, the wider diffusion and selection of ideas can be the most difficult point, whereas variation of ideas and opinions is more common. The reason for this is the tendency to homogenization rather that variation and selection in these institutions. But in public and social institutions there may sometimes also exist special opportunities for selection. This seems at least to be the case in the example, which is discussed in this chapter.

Opinions and ideas can come from employees as well as consumers and managers and from society more generally. There is an element of "voice" (Hirschman, 1970) in the caring approach to innovation: innovation is not only based in consumers' choices, but also in expressions of opinions about innovations and change. Innovation with care corresponds more or less to Karl Popper's idea of "piecemeal" social change as opposed to utopian social engineering, hence it is a critical-reflective and incremental approach to innovation (as explained in Popper, 1962).

A crucial aspect of innovation with care is therefore whether and how people achieve the relative freedom to independently form and express their opinions about innovations and fight for social values. This is not always obvious in public innovation, which can be characterized by isomorphism and other constraints (see below). On the other hand, public institutions (such as public schools) are often anchored in local communities, and the professionals working here have extensive autonomy to define and carry out their job.

Furthermore, as mentioned, this approach also cares about the ways in which opinions and ideas about innovations are collected and diffused through some kind of exchange and selection mechanism, assuming that some ideas are better, more relevant or more effective than others (see the ideas in Surowiecki, 2004). This point is crucial to this chapter and opposite to the homogenization thesis of institutional theory (DiMaggio and Powell, 1983). I try to demonstrate the critical role of certain "social and strategic arenas" in innovation with care in the case of the library, where variation and selection are combined in an imaginative way.

A social and strategic arena could be defined as a relatively robust place where actors from different settings can demonstrate ideas for each other, and where an imaginative selection and diffusion of workable ideas can take place.

MODES OF INNOVATION

Fuglsang and Sundbo have argued that it is useful to distinguish between different "modes of innovation" (Fuglsang and Sundbo, 2005). These modes define innovation at the systemic level rather than at the individual level. Such modes could be seen not just as theoretical abstractions; they are also regulative ideas for innovation in practice. Furthermore, they should not be seen as linked firmly to specific actors. Actors can move back and forth between them and position themselves in different ways at different times depending on the circumstances.

As a basis for this chapter, I will distinguish between four modes of innovation that are relevant to public innovation (see also Fuglsang, 2007). These modes of innovation are derived from the literature on innovation and institutional theory, but they can, as mentioned, also be seen as models that structure action in practice when people embark on innovation activities. These four modes are: entrepreneurial innovation, institutional innovation, open innovation and strategic reflexive innovation.

What follows is a brief review of these four modes and their relation to innovation with care. These modes are related to innovation with care in that they, to a greater or lesser extent, require a caring and imaginative approach to variation and selection. For example, open innovation requires a caring approach especially to selection, but also to the intermediation of selection and variation. Innovation with care means that initiatives are taken, at the systemic level, to provide for appropriate mechanisms of variation and selection.

Entrepreneurial Innovation

Entrepreneurial innovation is a mode of innovation which from a certain perspective does not seem to have much to do with innovation with care, because it is based in an individual person's ideas and opinions. But if we view entrepreneurship in a somewhat broader perspective as a societal activity that engages many individuals, entrepreneurial innovation can have a lot to do with innovation with care. By entrepreneurial innovation, I mostly mean the entrepreneur as described by Schumpeter in his book from 1911, published in the English language in 1934, on *The Theory of Economic Development* (Schumpeter, 1934, 1969). Chapter II of this book, entitled "The fundamental phenomenon on economic development," contains a fascinating story about the entrepreneur as a special "type" that has a special function for economic development. The function of the entrepreneur is to create new combinations of material and forces, that is innovation. Israel M. Kirzner has provided a somewhat different theory of entrepreneurship which is more consistent with economic theory (Kirzner, 1973). According to him, the entrepreneur is someone who corrects for market failures, especially when economic actors have bad information about prices. In that case, the entrepreneur can provide goods and services at the right price. In Schumpeter's approach to entrepreneurship, the entrepreneur is a dynamic force. In Kirzner's approach he is a stabilizing force that restores economic equilibrium. Schumpeter also describes the entrepreneur as a special type who has an inferior cultural position in society, and as a person who is driven by egoistic but not hedonistic motives.

At first sight this is, as mentioned, not innovation with care, but a dynamic and energetic person who breaks down established ways of doing things in a rather careless way – one should think. However, if we shift the perspective from the individual to the societal level, we can understand entrepreneurship as being structured at the systemic level, as a "mode of innovation." Entrepreneurship has to be recognized and approved of by society, and people who engage in entrepreneurship must somehow be able to see this as a career opportunity which is recognized and protected. We can add that no one, according to Schumpeter, is an entrepreneur throughout his whole life, and that all of us act as entrepreneurs from time to time. If we, in this way, can think of entrepreneurship at the systemic level, we can understand how entrepreneurship may fall under our definition of innovation with care: Entrepreneurship allows people to express their opinion in an independent way, and it works inside an economy or a society, where everybody from time to time can have the position of an entrepreneur.

A weak point of entrepreneurship as a socially structured phenomenon could be that a mechanism of selection is seemingly lacking. Entrepreneurs have themselves to pave the way for the new innovations and they must teach the consumers to use them, according to Schumpeter (1934, 1969: 65). As a matter of fact, this is an expression of innovation with care: the entrepreneur takes care of the diffusion and selection mechanisms. Entrepreneurship as a socially structured phenomenon can also lead to the formation of small kingdoms (p. 93) or cliques that are mutually in conflict with one another in an unproductive way, creating all kinds of market failures concerning price and quality. But entrepreneurs can also be thought of as people that work inside an established market economy, providing incremental innovations and improvements of goods and services. They take pro-active care of variation and selection.

Institutional Innovation

Institutional innovation is a mode of innovation that has an ambiguous relation to innovation with care. On the one hand, institutional innovation is usually understood as incremental and careful. On the other hand, variation and selection may sometimes be constrained by this mode of innovation.

To explain this mode of innovation, we can look at some insights from new institutionalism. In institutional theory, institutions are usually explained as stabilizing structures. Scott (1995: 33) defines institutions as follows: "Institutions consist of cognitive, normative and regulative structures and activities that provide stability and meaning to social behavior. Institutions are transported by various carriers – cultures, structures and routines – and they operate at multiple levels of jurisdiction." Given this definition, institutions are almost the opposite of innovation. On the other hand, as we have argued, innovation is also dependent on certain mechanisms of exchange that can lead to selection, that is certain institutional, stabilizing structures.

Furthermore, new institutional theorists have also sought to explain how change processes become institutionalized (Meyer and Rowan, 1977; DiMaggio and Powell, 1983; Røvik, 1998). In these discussions, the concepts of isomorphism and homogenization are critical. According to this approach, organizations change because they have to adapt to changing institutionalized environments in order to remain legitimate. These environments change because they come under pressure from popular "rationalized myths" (Meyer and Rowan, 1977) that demand more and better coordination in an increasingly complex society. These myths institutionalize in certain new government programs, educations and technologies. As such they put pressure on organizations to change. Institutional changes therefore lead organizations to become more similar (isomorphic). This mechanism is, however, a mechanism of homogenization rather than mechanism of variation and selection, and it is grounded on legitimacy rather than relevance or effectiveness. Homogenization can also lead to a decoupling effect where organizations pretend to follow the external requirements while in practice they develop in a different direction. Some institutional theorists have struggled to work out a more nuanced approach to institutional change, adopting the concept of "translation" in order to describe how organizations in practice translate the institutionalized ideas into something relevant at the local level (for an overview see Scheuer, 2006).

Isomorphism and other constraints on variation and selection are not really consistent with innovation with care. The reason for this is that innovation with care cares about how people can independently and freely form and express an opinion and can fight openly for social value. And it cares about selection rather than homogenization. This is not really the case under conditions of institutional innovation as described by some of these theories: either people seemingly give in to institutional pressures and become more similar for reasons of legitimacy rather than selection, or they protect and blur their opinions behind a facade of change. Nevertheless, isomorphism, where actors orient themselves toward the same organizational fields, is also to some degree a precondition for innovation with care. The expression of opinion and a critical reflective approach to change and innovation assumes that people are members of a community where these opinions can be exchanged, explained, discussed and selected, and where structures exist in which certain changes can gain acceptance. Thus, institutional innovation can be seen as both a precondition of, and a barrier to innovation with care. Institutional innovation can constrain variety and selection due to homogenization processes. But it also provides a framework for communication and selection of opinions.

Open Innovation

Open innovation is a mode of innovation which has been widely examined and discussed over recent years in the literature on innovation based in Chesbrough's original explanation and definition of this phenomenon (Chesbrough, 2003). Also von Hippel's work on the democratization of innovation could be said to belong to this approach (von Hippel, 2005). Open innovation is a historical approach to innovation that becomes relevant because of changing precondition for innovation. Chesbrough defines open innovation, as opposed to closed innovation, in the following way: "Open innovation is a paradigm that assumes that firms can and should use external ideas as well as internal ideas, and internal and external paths to market, as the firms look to advance their technology. Open innovation combines internal and external ideas into architectures and systems whose requirements are defined by a business model. The business model utilizes both external and internal ideas to create value, while defining internal mechanisms to claim some portion of that value" (Chesbrough, 2003: xxiv). Hence, firms and organizations must scan the environment for ideas, perhaps through assistance from other economic actors, and try to figure out which ideas to adopt and which to let go. Companies must try to capture some value from these external ideas through internal processing of them. The background for the transition from closed to open innovation is, according to Chesbrough, that innovative resources have become much more widely distributed in society than just a few decades ago. Innovative resources are no longer just concentrated in firms' R&D laboratories or in government labs and universities. They are distributed to a wide number of diverse actors such as new spin-offs from old companies, new knowledge intensive services, groups of advanced users, clever users equipped with powerful computers, customers with concrete new ideas, and many more. No longer can firms control their development resources inside-out or deal in a controlled way with the environment as in a chess game. The game becomes, according to Chesbrough, one of poker, where you can attempt to "read" the opponents, stay in the game as long as possible, or fold your cards if they are not good enough.

Open innovation is in some ways consistent with innovation with care. Obviously, a variety of opinions and ideas are involved. It is not enough to rely on internal planning or legitimate ideas. Previously in this book we have argued that open innovation can be seen as a part of a Schumpeter III approach to innovation characterized by a growing fragmentation and distribution of the means of innovation (see the introduction to this book). Companies and institutions are trying to capture value from these resources and translate them into something useful at the organizational level or letting them go if they have no obvious function for them. On one hand this can lead to a multiple opinion approach, but on the other hand this may also mean that the goals and the rules of the game are becoming more blurred, and therefore opinions become more difficult to select and recognize. As in poker, this can lead to a positioning game where people try to conceal rather than reveal their "ideas" (see also Davies and Davies, this volume).

At the systemic level, understood as a socially structured form of behavior, open innovation may lead people to collect opinions from others, while at the same time concealing some of their own. For different and more complex reasons than in the case of entrepreneurial innovation, this approach therefore seems to lack a proper mechanism of selection. In other words, the "pokerness" of open innovation could be a main problem, a self-destructive approach. A critical requirement of open innovation could therefore be a more caring approach to selection. For one thing, open innovation, contrary to poker, in practice probably requires a good deal of trust among actors.

Helga Nowotny and others have tried to use the metaphor of "agora" to understand what goes on when science become more open, and this may also be a good description of the requirements of open innovation beyond "pokerness" towards innovation with care.

> Going back to an old Greek term, we call it the agora. It requires the management of complexity in a public space, which is neither state, nor market, neither public, nor private, but all of this in different configurations. Indeed, the agora is everywhere. It is in your mind as much as in social or public political settings, in corporate structures or in the rules of governance as much as inside laboratories and how we relate to each other. It still recognizes disciplines, but it has moved beyond them to engage with – whom? – the imaginary layperson and imagined users, the public, citizens, in short, what we take to be society to whom we all belong. (http://www.interdisciplines.org/interdisciplinarity/papers/5)

The metaphor of agora points in the direction of the need for social and strategic arenas and platforms where problems can be discussed, opinions exchanged and ideas selected. This seems consistent with the discussion of strategic platforms in Hansen and Serin (this volume). In this case, the market is not the only mechanism of variation and selection. The caring approach is not simply based on supply and demand, but in interpretations of demand, and these interpretations need to be investigated, discussed and confirmed in the agora to some extent.

Strategic Reflexivity

The concept of strategic reflexivity is an attempt to go one step further compared to Chesbrough (Sundbo and Fuglsang, 2002; Fuglsang and Sundbo, 2005). In this approach, the organization is an interpretative system (see also Daft and Weick, 1984) that seeks to open up and interpret what goes on in the market and the environment as well as in the organization. Hence, this approach deals with the "interpretations." These interpretations lead the company to formulate certain strategies that can be used to intervene in organizational practices in order to create the architectures described by Chesbrough between external and internal ideas and capture some value from these. Hence, strategies are the mechanisms of selection, but they are based in interpretations rather than only in market pull. These strategies are also reflexive, because people are invited to respond to them. In this way, a variety of opinions are involved, but still there may, in practice, be severe problems with strategic positioning and people concealing their true opinions and social values.

A main problem in this approach is that there is a very thin line between social engineering on one hand and innovation with care on the other. Strategy-formulation is probably something that has to take place at the top-management level. Under conditions of open innovation strategy-formulation may often become part of a top-strategic approach in a concern-like structure. Clearly there is a risk of a gap between strategy and reflexivity. People may want to contribute, but how are their concrete ideas reflected in the overall strategy? In the end, strategy-formulation may lead to socially engineered but socially ineffective selection of ideas.

To conclude, we have reviewed four modes of innovation. They structure the process of innovation in different ways and they assign different roles to innovation with care, that is caring for variation and selection in a proactive way. In the entrepreneurial mode, a variety of opinions and ideas can be mobilized, and the entrepreneur must teach the consumer to use innovations in a selective way. In institutional innovation, the main mechanism of organizational change is homogenization rather than variation and selection. In open innovation and strategic reflexivity, variation and selection potentially play important roles. Here it becomes crucial for organizations to care about multiple opinions and seek to integrate them into a strategic perspective. But there remains a risk of fragmentation between strategy and reflexivity. I argue that social and strategic arenas, or agoras, where opinions can be exchanged, integrated and selected could therefore be crucial to variation and selection in open innovation and strategic reflexivity.

In summary, the contention of this chapter is: (a) that open innovation and strategic reflexivity have become more important as a mode of innovation; (b) that open innovation and strategic reflexivity require care for variation and selection; and (c) that social and strategic arenas where opinions can be exchanged and integrated can be an expression of this at the systemic level.

In the following we examine how innovation with care works in the Danish public library. One thing we show is how the mode of innovation is changing from institutional to open and strategic reflexive innovation with a growing emphasis on variation and selection. Furthermore, we examine how social and strategic arenas are constructed, and can provide frameworks for variation and selection.

From the case study I try to develop the notion that the library as a social institution is actually quite fit for open innovation and strategic reflexivity as well as innovation with care. The library is tied to certain social and political structures that can lead actors to orient themselves toward certain social and strategic arenas inside the library system.

THE CASE OF PUBLIC LIBRARIES

Over the past decade, several innovations have been launched in the domain of the Danish public library and these bear witness to the growing impact of open innovation as well as the complementary changes in the field of innovation toward the formation of new social and strategic arenas as defined above (a relatively robust place where actors can demonstrate ideas for each other, and where an imaginative selection and diffusion of workable ideas can take place). This means that attention is increasingly paid to mechanisms of innovation, variation and selection rather than to processes of homogenization.

First, in the field of innovation, the overall organizational framework of the library sector has been organized in a new way. New social and strategic arenas are being created that are enabling better selection and diffusion of ideas. For example, The Danish National Library Authority plays a more arena-like role for library development. The Danish Bibliographic Centre (DBC), which is a private limited company, has also gained a more critical role for library development and selection.

Second, new e-based self-services have been introduced that can reach out to library users in new ways. They are created both through contracts between the Library Authority and DBC, and by libraries collaborating together. They can apply to the Library Authority for funding. Some of these services, such as bibliotek.dk (see p. 102) become, in themselves, social and strategic arenas for variation and selection.

Third, the librarians' roles have been discussed and are interpreted in new ways as "information specialists and cultural intermediaries" (subtitle of the librarians' union). For example, the library school, which is another social and strategic arena for variation and selection, has become a "Royal School of Library and Information Science." The library education that is carried out at this school has been turned into a modern media education.

Fourth, several initiatives have been taken to change the architectures of the library room so as to become more lively and experience-based. Kolding library provides an interesting example, with "libraries-in-the-library" that allows, for example, for noisy activities to be separated from quieter ones.

And fifth, a range of new digital tools and materials has been introduced into the library (PCs, Internet access including special services, CDs, DVDs, PC games and so on). Today, all public libraries offer Internet access to the public.

These changes may not be so impressive from a technical point of view, but from an organizational point of view they are. The library is being changed into an open and dynamic environment that is supposed to play a mediating role between the information society and the citizens. It presents itself as a gateway and guide for people to the many emerging information resources on the Internet. The image of the library is also changing from being a quiet and partly boring place to becoming a livelier place. In a specialist journal for the Danish library sector, the librarian is seen as someone who is "proactive, extrovert and ready for changes and is good at selling herself and create her own job" (*DBCAvisen* 04 Winter 2005). Therefore, "librarians today are much more than an elderly lady with an amber chain that wears her hair in a knot at the back sitting behind the counter of the public library" (*DBCAvisen* op. cit.).

The director of the Danish National Library Authority argues that the library of the future could be seen as a factor for competitiveness for Danish firms. "(T)he issue is now how to develop the libraries into becoming extrovert knowledge networks that can help citizens and especially companies with knowledge whenever they need it . . . The goal is that libraries in the future provide easy access to relevant information in the work situation just like waterworks provide clean water in the tap" (*Mandag Morgen* No. 3, 23 January, 2006).

These changes of the library image can be seen as a case of open innovation in the following sense: On the one hand, competition from search engines such as Google or Yahoo, and new possibilities to download information and entertainment from the Internet have forced the library to be more aware of external developments. Libraries have acknowledged that they have to reach out to library users in new ways in order to convince them that they still need the library. To do so, the library must try to capture values from external ideas, hence the many new information and networking technologies, and combine this with internal resources and ideas.

At the same time, several new social arenas are created, as indicated, where actors can demonstrate ideas for each other and recognize and select the ideas that work for them. They include the restructured Library Authority and DBC, some of the new e-based services, the reformed Library School including the ongoing teaching and research, the librarians' union and some of the library associations (see p. 100).

For the library to transform itself along these lines, a field of innovation is needed where many actors are involved in deliberations about the future of the library. This field is constituted by a combination of political pressure, professional awareness and social orientation. For example, many of the new challenges and opportunities for the new library were explored in a report from a committee under the Ministry of Culture, the *White Paper on Libraries in the Information Society* (Udvalget om bibliotekerne i informationssamfundet 1997). This report led to a new Library act in 2000. Here, the purpose of public libraries is explained as follows:

> The purpose of public libraries is to promote enlightenment, education and cultural activity by providing books, journals, audio books and other relevant materials such as carriers of music and electronic information resources, such as the Internet and multimedia. (Lov, nr 2000)

The public library is still defined as a social institution that primarily promotes education. But in practice, huge pressure is now put on the library to transform itself. No longer should the library just deal with education, but also with the information society and many more related issues such as the continuous need for competence development of citizens and the workforce in an information and knowledge society, or the training of elderly people to be able to participate more in the Internet-based society.

The creation of the open and lively library is also seen as a new collective challenge for the libraries. Dyrbye argues that the Danish libraries during the 1970s and 1980s became more individualized and fragmented (see Dyrbye *et al.*, 2005: 9–37). Each library could have a tendency to conduct its own policy, and, during the 1980s, entertainment came to play a more important role for many libraries. Variety was the rule. But from the mid-1990s onwards, due to the advent of the Internet, there was a renewed interest in the library cause as a joint project for the library sector in Denmark (ibid.: 37).

In summary, the library becomes an open and lively environment that mediates between the individual user and the information society, leading to simultaneous changes in the domain of innovation (the library), the mode of innovation (towards open and strategic reflexive innovation), and the field of innovation (the formation of new social and strategic arenas). In order to create a framework for innovation, variation and selection, the libraries and librarians start to orient themselves more toward certain social and strategic arenas.

THE LIBRARY AS A SOCIAL INSTITUTION

In the following I argue that the "social institution" of the library sustains the creation of social and strategic arenas in the library sector. As a social institution, the public library has a potential to create certain socially effective arenas of variation and selection. This arena orientation was not so important in the 1970s and 1980s when the Danish libraries were more fragmented. Today, the field of innovation has become more selective, and therefore the complex social structures that lie underneath the library sector become relevant for the mobilization and the arena-orientation of librarians.

Danish public libraries are clearly associated with a social movement called the library cause. The library cause was especially active in the beginning of the twentieth century trying to convince political decision makers about the necessity of libraries for Danish society. The movement was inspired both by the European enlightenment movement, similar movements in other countries, and the development of the Danish democracy and welfare society. The library cause saw the role of libraries in the context of education and democracy (see especially Lange, 1909). Knowledge promoted through libraries and education was seen as a precondition for a democratic society (see Lange op. cit.: 5).

Today, the library cause has been institutionalized in many ways, and it is therefore associated with a number of organizations and societies.

The Library Association is one such organization which fights for "the free, open, democratic society with free and equal access to information and culture for all" and for "the defence of democracy by making public libraries and the library sector as a whole an important actor that can help ensure that the whole of the population has free and equal access to all information" (from the Library Association's Strategy, see www.dbf.dk/Default.aspx?ID=2803).

The Library Association, which was formed more than 100 years ago, is an interest or lobby organization, which is organized into five regional library societies. Members are the municipalities, represented by politicians who specialize in cultural policy. The association also counts about 500 personal members (librarians) and about 70 institutional members.

The Danish Union of Librarians is another important organization which promotes libraries. The union presents itself on its home page as "a union of information specialists and cultural intermediaries." The union "has about 5500 members of which 2600 are employed by municipalities, 900 by the state and 500 by private firms." In recent years, the union has actively tried to reformulate both its own role and that of the librarian. Its chairman is, for example, quoted for giving strong support to user driven innovation on its homepage on 10 January 2005.

A number of other small societies can be identified that are working more or less in accordance with the library cause, for example the Library Leaders' Association (Bibliotekslederforeningen), The Association of Danish Research Librarians (Danmarks Forskningsbiblioteksforening), The Association of Danish School Librarians (Danmarks Skolebibliotekarforening) and Municipalities' School Library Association (Kommunernes skolebiblioteksforening). An association also exists that pulls together the different strings, called the Library Umbrella (Biblioteksparaplyen), which deals with subjects of general interest in the Danish library sector.

Thus, one finds a myriad of different societies supporting the library cause. A very visible and lively discussion is going on in these settings, including pamphlets, books and magazines published by the agencies, the companies (DBC) and associations that bring into focus both the value and spirit of library work and the idea that it is necessary to change certain things.

What are the implications of this? First and foremost, what we can notice is that there are obviously some interdependencies here between the library cause and library development, in this case the attempt of the library sector to capture values from external ideas. The librarians and the libraries as well as the municipalities and the library union are mobilized for a new cause, not just for survival and legitimacy. This is in itself an important factor.

Moreover, as we have said, the social structure of the libraries also makes it possible for people to orient themselves more easily toward certain social and strategic arenas of library development. Even if the individual libraries may have different cultures, there is a "field of innovation" within which there already is a culture of deliberation, exchange of opinion, meetings and so on, which makes it easier for librarians to communicate across individual libraries. An additional factor that also underpins this strategic field of innovation is the Library School where librarians are educated. Education is an important mechanism of exchange, variation and selection.

INNOVATION OPPORTUNITIES

Two examples of key innovations in the Danish public library in recent decades are the danbibbase from 1993–4 and bibliotek.dk from 2000. Another important example of a recent innovation is the Library Watch from 1999, an information service created by several collaborating libraries. These three innovations demonstrate the simultaneous changes that have been taking place, as discussed above, in the domain of innovation, the mode of innovation and the field of innovation.

The following brief account of these three innovations is based on interviews which were conducted in 2005 with a vice-director of the Danish National Library Authority (VD), with a head of department in the Danish Bibliographic Centre (HD) and a project initiator and project developer (PD) of the Library Watch.

The danbibbase was developed by the Danish Bibliographic Centre (DBC) as one of this company's first and largest tasks after its restructuring to become a private limited company in the beginning of the 1990s (see Nielsen *et al.*, 2005: 490ff). In international comparison, the danbibbase is, according to VD and HD, a rather unique database covering all Danish libraries' materials. It exploits the circumstance that Danish libraries all use the same catalogue system and all have developed their own electronic databases over the years. The Danbibbase also contains a national bibliography and it communicates daily with an American database in order to list as many publications as possible from all over the world. The database is continuously updated every time the individual libraries purchase new material. In this way, it is possible immediately to see what materials the individual libraries have acquired.

According to HD, it took several steps to develop the danbibbase. The first step was taken by the individual libraries. During recent decades, they have developed and maintained their own electronic databases registering their own stock of materials. Private firms have often been involved in developing the different local systems. The process of developing the danbibbase was, however, a more top-strategic effort with DBC as a leading force. The danbibbase draws directly on the many databases that have been locally created. This is possible due to some of the characteristics that are inherent to the domain of the public library: librarians are used to catalogue work and to work with common standards.

Every time a library acquires new material, the local librarians will register this material in their own local system. If the material is already mentioned in the danbibbase (as registered in other libraries, the national bibliography or the US Library of Congress), the librarian can use this information for local registration. Every day all libraries then send files to DBC with a list of the new materials that have been acquired. This is then uploaded in the danbibbase. In this way, it is possible for all libraries very quickly to identify what libraries have acquired which specific material.

In this way, the domain of the library is changing from a physical to a virtual library. Furthermore, the mode of innovation is also changing: The construction of the danbibbase relies both on external ideas (the use of information technology) and internal ideas (the librarians' ability to work with common standards). In addition to this, the broader field of innovation is also being transformed. The danbibbase and DBC become new social and strategic arenas for library work where librarians can exchange information about library development and see what others are doing.

Users are also increasingly included in the domain and field of library development. An example of this is bibliotek.dk, the users' Internet access to the danbibbase.

When the Danish Parliament passed the library act in 2000, means were set aside in the national budget to finance the development of Internet access to the danbibbase (Pedersen *et al.*, 2004). Bibliotek.dk exploits the previous construction of and continuous development of the danbibbase with certain exceptions, and thereby it brings library work much closer to the citizens.

Over the years, Bibliotek.dk has become a flagship for the new library cause and is an eye-opener for libraries and library users. The users now have entirely new possibilities for browsing the library and order materials. And the libraries have had entirely new possibilities for developing new e-services on top of bibliotek.dk, such as the Library Watch (see below).

Library users can browse and order materials from a home PC using bibliotek.dk. All Danish libraries including the research libraries can be used for this, hence the user has gained access to a very large nation-wide "hybrid" library. Once an item has been ordered, the user can always collect the materials at the local library and will be notified by email. To make this possible, a car company has been hired to transport the books between the libraries in the country. About 1.4 million items were exchanged among libraries in one year (2005). If a book is not available in the libraries, bibliotek.dk will refer the user to a bookstore where the material can be bought.

Bibliotek.dk is connected to a measurement made by Gallup about visits to websites. It shows that the number of visitors have increased by about 50 percent from 2004 to 2005. The number of orders has also increased. In 2003, there were 984 052 orders, in 2004 1 212 322 orders and in 2005 1 383 016 orders. Forty percent of the visitors were students and 33 percent from the Copenhagen area according to a 2004 user survey (http://bs.dk/content. aspx?itemguid={DB69FFAE-12B1-4D43-9026-C9E3A1A381F1}).

DBC receives, according to HD, a number of mails from users, and 9000 users are connected to a newsletter and receive questionnaires in connection with DBC's and the Library Authority's development plan. According to HD, users are very good at giving feedback and can also sometimes come up with new ideas.

In September 2005, a new look was introduced for bibliotek.dk. The style was modernized, but, according to HD, the homepage still had to signal credibility, neutrality and seriousness. Therefore, animations are excluded. The reason for this was a user survey showing that library users wanted a traditional library look without strong colours.

A number of new functions have been added to the homepage. There is now a site were users can click their way through to certain subjects similar to Yahoo's homepage. Many library users have, according to HD, expressed a very positive attitude to this innovation. The purpose is to make it easier for users to find quality literature. Certain subjects, for example "suicide", give rise to so many hits that users cannot identify the relevant literature, which perhaps is ranked as number 5000. Apart from that, people who have problems with spelling can have difficulties in using the search functions, which requires correct spelling. In this way, bibliotek.dk functions as a guide for the user.

In a similar vein, the database also makes distinctions between different editions of a work, for example the opera *Aida* on CD of which there is a huge number of different versions. This part of the project has taken about five years to develop. An advanced search function has also been added for those who do not know the search codes or do not like to use such codes.

HD concludes: "The Internet has meant democratization. Before the advent of the Internet, citizens had to ask a librarian. Now, they can do everything by themselves. Something has really happened. Clearly, there is another purpose also of the digitalization, namely to rationalize the working routines. Libraries have not had more personnel. Rather, the opposite is the case."

The development of bibliotek.dk has taken place in a concern-like structure with the Library Authority as a social and strategic arena of selection. The starting point is certain values that have been defined by the Library Authority and a development plan, which has been created by DBC and has been submitted to the libraries for critique. It is the Library Authority that makes the final decisions, but in practice, the responsible person in DBC has worked out the plan (HD). Hence, this is a strategic reflexive approach to innovation.

The input to the development plan also comes from users, libraries and from other search engines, for example Google. In December 2004, DBC worked out a proposal for the new homepage, its look and functionality. After that, a contract was set up with the Library Authority. According to HD, there was close cooperation among the parties. The development plan was then submitted to the libraries and a decision was taken in December 2004. Hence, bibliotek.dk is emerging through a kind of innovation partnership between the Library Authority and DBC even if the Library Authority is commissioning the work.

The innovation process is one of open innovation. In the first phase of the project, the focus was on the design of the homepage and HD contacted a design company. She hired an external consultant who could help her with the specifications and give proposals for improvements. The company was hired to design the look of the homepage.

Google including Google scholar and Google book search have also been important to this open innovation, and bibliotek.dk is trying to capture some of Google's value while also dealing strategically with Google. The strategy is to cooperate with Google so that the two services can complement each another.

A project has been developed between Google and bibliotek.dk, which since late 2006 has made it possible to gain access to certain catalogue services of Danish libraries from Google Scholar and Google Book Search. The attitude toward Google is quite clear: Google is far from perfect, but the libraries will have to live with the fact that users think that Google is almost perfect.

We can conclude that bibliotek.dk has developed as a project, which reflects changes in the mode of innovation, the innovation domain, and the field of innovation. The mode of innovation is becoming one of open innovation. Rather than defining the value of the library inside-out, the library is on the alert for external values that can be captured and integrated with internal values. This in turn changes the domain of the library, the library becomes more extrovert and sensitive to variation and selection. Furthermore, the role of the library becomes that of a guide in the information society. Finally, the field of the library is also changing. The field is no longer the local library and community. The field is, in this case, the social and strategic arenas of the Library Authority, and bibliotek.dk is becoming a social and strategic arena where multiple opinions are integrated and truly selected.

While the danbibbase is developed by DBC in close cooperation with the Library Authority, the Library Watch is an example of a service which has been developed from the bottom by single entrepreneurial libraries in innovation partnerships, and then financed by the Library Authority. The Library Watch is linked to bibliotek.dk (on its home page) together with other "netlibraries," which are other specialized services that are produced or co-produced by single libraries.

The Library Watch is an information service where the user can ask questions concerning books, school work or hobby. This is done by filling in a form at the homepage of the Library Watch, which is sent to a librarian, or by chatting with a librarian in a chat room during the opening hours of the Library Watch. Currently (March 2007), 45 public libraries and 16 research libraries and about 250 librarians cooperate on this service.

According to PD, the librarians working with the Library Watch choose to see it as an extension of the librarians' work domain, "25 years ago, the job had a very different content," says PD. "If the user could not come up with a very precise question, the librarian could not give an answer. Today, one almost never gives up."

A typical use of this service could be a student in primary or secondary school that needs information and literature for an assignment in school. The student asks the Library Watch by filling in the electronic form and will receive answers within 24 hours. For example, the student may have an assignment about human rights. The librarian will then explore this issue and mail references to the student which could include links to important websites as well as materials in the library. The librarian will typically also comment upon the materials that are referred to. If this goes on in a chat room, then the student may ask for further references.

The Library Watch is a good example of changes that have taken place in the innovation domain transforming the librarian into a guide in the information society. But it is also an example of a social and strategic arena and of open innovation: It constitutes a social and strategic arena where librarians who participate in the service can make use of each other's work. For example, all answers that are given to users in the service are kept in a database, and these can be browsed by the librarians. Furthermore, it is an innovation which is strategically selected and recognized at the higher level of bibliotek.dk and the Library Authority.

ANALYSIS

The above cases are significant achievements of the Danish library sector. They demonstrate how a new concern-like structure has been established around the Library Authority in the library sector that can stimulate variation and selection rather than homogenization. This "concern" has different characteristics.

1. It is domain specific, which means that it is concerned with library work and not other services. Nevertheless, it pulls together different activities, such as public libraries, research libraries, library education and development of e-based services, around certain social and strategic arenas, such as the Library Authority itself or bibliotek.dk.
2. It enables innovation partnerships among libraries. These innovation partnerships can evolve around single entrepreneurial libraries, or they can involve all libraries as well as DBC and private consultants. This creates variation.
3. There is a growing emphasis on the selection, diffusion and scaling up of innovations. If previously the single libraries had freedom to use their own ideas for library development, today such ideas and single experiences are recognized and shared to a larger extent and scaled up as in the case of the Library Watch, where more and more libraries are invited to participate, and experiences are exchanged across librarians from different libraries.
4. This also implies that library development is viewed more and more from the systemic level. While on the one hand libraries have freedom to experiment with their ideas, they are not acting on their own or on behalf of a single library, but on behalf of the whole library sector as such.

We have seen how the mode of innovation is changing toward open innovation and strategic reflexive innovation. The library must seek to capitalize external ideas within its own strategic reflexive framework. This also leads to changes in the domain of innovation (the librarian becomes a guide) and in the field of innovation (the creation of the concern-like structure and new social and strategic arenas for library development). But the new library also incorporates features from the two other modes of innovation.

Institutional Innovation

This is still an important starting point for the Danish library sector. References are made in many places to the library cause as a basic framework for innovation, and the roles and functions of people working in the library sector are highly institutionalized with a common framework for education and many internal debates and discussions going on about the future of the library. While the image of the librarian is being changed, it is still the basic skills of the librarians in the basic library domain that are in focus. In a certain sense, the advent of the Internet makes it possible to promote the library skills for new purposes.

But the social institution of the library is also challenged in several related ways. For one thing, there is a "journey to the interface" going on (see Parker and Heapy, 2006). This means that the interface between the librarian and the library user becomes more important as a complement to internal standards and quality criteria. Furthermore, libraries and librarians have acquired freedom to propose projects. In addition to this, libraries and librarians are, as mentioned, increasingly supposed to act on behalf of the library as a system and to upscale new experiences and ideas to the system level rather than orient themselves towards local practices and routines. Finally, the library becomes more proactively involved in changes rather than reacting to them, they are shaping the answers rather than merely detecting them from the external environment (see Daft and Weick, 1984). And they are involved in processes of variation and selection rather than homogenization.

Entrepreneurial Innovation

Initiatives such as the Library Watch and other net libraries are to a large extent dependent on entrepreneurial initiative, whereby innovation opportunities are identified by a small group of librarians and where the projects become managed bottom-up by the librarians themselves. Entrepreneurial types who want to create something new or correct perceived deficiencies in the existing library system are given space for action. This is evidently becoming a career pattern in the library sector, where you can increasingly do new things with funding achieved from the Library Authority.

But as we have seen, entrepreneurship is not working entirely in the Schumpeterian way. In the case of the library, we must speak of entrepreneurship as a collaborative effort, where many entrepreneurs are working together in the overall context of the library system and the new social and strategic arenas. Moreover, the entrepreneurial initiatives such as the Library Watch themselves help create strategic fields for innovations where many different opinions and ideas can be tied together and tested against each other.

Open Innovation

What especially seems to matter a great deal for current developments in the library is, however, an emerging pattern of open innovation. This is due first and foremost to the advent of the Internet and search engines such as Google that constitute external threats to the library. This has forced the Danish library sector to watch out for the values that users generate on the Internet, and thereby to enter into a new mode of innovation where the ability to capture external ideas and values play a growing role.

Strategic Reflexivity

Strategic reflexivity is reflected in the way in which the Library Authority and DBC are being transformed into top-strategic organizations, and in the strategy discussions going on in the library sector as a whole. There is a continuous search for effective interpretations and key problems that can inform a new library cause, common strategies that in an effective way can ensure the role of the libraries in the future. These strategies are at the same time reflexive, meaning that they link with the ideas and opinions of staffs and users. Open innovation and strategic reflexivity are really two sides of the same coin.

Open innovation and strategic reflexive innovation are, as we have argued, intertwined with innovation with care. Innovation with care is an approach to innovation that cares for variation and selection and their combinations. This means on the one hand that the freedom to express opinions and ideas are important and that mechanisms for recognition and selection of them are crucial. This is possible, we have argued, because the social institution of the library can potentially generate social and strategic arenas to which people in practice will be willing to devote attention.

The practical consequences of the approach are that there is a difficult balance to maintain between the requirements for openness and open innovation and the request for social institutions that in an imaginative way can ensure interactions and recognitions of innovation and provide the necessary social glue, between variety and selection.

CONCLUSIONS

The public sector is often understood as bureaucratic and static and the literature on change processes in the public sector often emphasizes homogenization, isomorphism and decoupling effects rather than innovation, variation and selection.

In this chapter, I have discussed how the public sector can also be seen as a context that can comfort innovation, variation and selection. A number of factors for that have been discovered using public libraries as a case.

1. Innovation has always been important to the library sector in the form of institutional innovation – closely linked to the social movement called the library cause.
2. Competition has become a critical factor for the library sector almost as a paradigm shift from welfare state to competition, even if the welfare state still supports libraries. This also means that external ideas of innovations and attempts to capture value from these have a much greater impact on the library than before, and a framework of "open innovation" based partly in external ideas clearly becomes more pertinent to innovation planning. This represents a change in the mode of innovation.
3. As a response to the growing competition and openness, the libraries seek to identify new societal problems that can target and inspire the innovation process. There is a shift from an emphasis mostly on democracy and enlightenment (the old library cause) toward an accent also on citizens' development of competences in the knowledge society (the new library cause). The domain of innovation (the library work) is being transformed: the librarian becomes the citizens' guide in the information society.
4. Another very critical factor has been the careful construction of concrete social and strategic arenas were opinions can be exchanged and selected – this represents a new field of innovation. This has been done through a process where politicians, library experts and the library movement have all been involved in deliberations about the library cause, and the public library has been organized in a concern-like structure. These arenas provide a tool for exchange of opinions and ideas across libraries, and recognition and selection of these opinions and ideas.

In all this, it appears, as indicated, as if the mode of innovation is changing, from institutional to open and strategic reflexive innovation. Or rather, the library is constituting itself as an open and strategic reflexive innovation system. This means that attention is paid to variation and selection as mechanisms of organizational change rather than to homogenization.

We have suggested that innovation with care is an underlying mechanism of all this: the library actors need to orient themselves carefully and imaginatively toward mechanisms of variation and selection. I have argued that they appear to do so to a large extent. One explanation of this may be found in the social institution of the library and the library cause, which lies underneath the library sector as a social glue.

Even if innovation, variation and selection seem to become important mechanisms of organizational change in the Danish library system, homogenization and institutional innovation may still prevail in many cases of public innovation. For future research it may be interesting to examine the shifts that can take place between variation and selection on the one hand and homogenization on the other.

REFERENCES

Chesbrough, H.W. (2003), *Open Innovation: The New Imperative for Creating and Profiting from Technology*, Boston, MA: Harvard Business School Press.

Csikszentmihalyi, M. (1996), *Creativity: Flow and the Psychology of Discovery and Invention*, New York: HarperCollins.

Daft, R.L. and K.E. Weick (1984), "Toward a model of organizations as interpretration systems," *Academy of Management Review*, **9** (2), 284–95.

DiMaggio, P.J. and W.W. Powell (1983), "The Iron Cage revisited: institutional isomorphism and collective rationality in organizational fields," *American Sociological Review*, **48** (2), 147–60.

Dyrbye, M., J. Svane-Mikkelsen, L. Lørring and A. Ørum (2005), *Det stærke folkebibliotek: 100 år med Danmarks Biblioteksforening*, Copenhagen: Danmarks Biblioteksforening og Danmarks Biblioteksskole.

Fuglsang, L. (2007), "Interpreting the means and the goals: innovation in Danish public libraries," in S. Scheuer and J.D. Scheuer (eds), *The Anatomy of Change – a Neo-institutionalist Perspective*, Copenhagen: Copenhagen Business School Press (forthcoming).

Fuglsang, L. and J. Sundbo (2005), "The organizational innovation system: three modes," *Journal of Change Management*, **5** (3), 329–44.

Hippel, E. Von (2005), *Democratizing Innovation*, Cambridge, MA: MIT Press.

Hirschman, A.O. (1970), *Exit, Voice, and Loyalty: Responses to Decline in Firms, Organizations, and States*, Cambridge, MA: Harvard University Press.

Kirzner, I.M. (1973), *Competition and Entrepreneurship*, Chicago, IL: The University of Chicago Press.

Lange, H.O. (1909), *Bibliotekssagen uden for København. Foredrag på biblioteksmødet i Aarhus d. 4. august 1909*, Copenhagen: Dansk Biblioteksforening.

Lov (Act) no. 340 (2000), *Lov om biblioteksvirksomhed*, 17 May.
Meyer, J.W. and B. Rowan (1977), "Institutionalized organizations: formal structure as myth and ceremony," *American Journal of Sociology*, **83** (2), 340–63.
Nelson, R.R. and S.G. Winter (1977), "In search of useful theory of innovation," *Research Policy*, **6** (1), 36–76.
Nielsen, E.K., N. Chr. Nielsen, St. B. Larsen and S.L. Bak (eds) (2005), *Kommunikation erstatter transport: den digitale revolution i danske forskningsbiblioteker 1980–2005. Festskrift til Karl Krarup*, Copenhagen: Det Kongelige Bibliotek.
Parker, S. and J. Heapy (2006), *The Journey to the Interface*, London: Demos.
Pedersen, A.M.R., M.T. Lynge, H.K. Mikkelsen, C.V. Taankvist, B. Larsen, J.T. Hansen and J.H. Larsen (2004), *Udviklingen i folkebibliotekernes virksomhed og økonomi 1994–2002 med særligt henblik på evaluering af Lov om biblioteksvirksomhed (2000)*, Copenhagen: Biblioteksstyrelsen.
Popper, K.R. (1962), *The Open Society and its Enemies*, London: Routledge & Kegan Paul.
Rogers, E. (1995), *Diffusion of Innovations*, New York: Free Press.
Røvik, K.A. (1998), *Moderne organisasjoner: trender i organisasjonstenkningen ved tusenårsskiftet*, Oslo: Fagbokforlaget.
Scheuer, J.D. (2006), "Om oversættelse af oversættelsesbegrebet. En analyse af de skandinaviske ny-institutionalisters oversættelse af oversættelsesbegrebet," *Nordiske OrganisasjonsStudier*, **8** (4), 3–40.
Schumpeter, J.A. (1934, 1969), *The Theory of Economic Development*, Oxford: Oxford University Press.
Scott, W.R. (1995), *Institutions and Organizations*, Thousand Oaks, CA: Sage.
Sundbo, J. and L. Fuglsang (eds) (2002), *Innovation as Strategic Reflexivity*, London: Routledge.
Surowiecki, J. (2004), *The Wisdom of Crowds: Why the Many are Smarter than the Few and How Collective Wisdom Shapes Business, Economics, Societies, and Nations*, London: Little Brown.
Udvalget om bibliotekerne i informationssamfundet (1997), "Betænkning om bibliotekerne i informationssamfundet," *Betænkning no. 1347*, Copenhagen: Kulturministeriet.
Young Foundation (2006), *Social Silicon Valleys: A Manifesto for Social Innovation, What it is, Why it Matters and How it can be Accelerated*, London: Young Foundation.

6. Getting waste to become taste: from the planning of innovation to innovation planning

Gestur Hovgaard

INTRODUCTION

This chapter presents a single case study of the evolution of the Danish food-ingredient company Danmark Protein (DP). Danmark Protein, which today is fully incorporated with the dairy giant Arla, is a world leading producer of highly sophisticated and biotechnologically derived whey proteins. There are at least three major reasons why this company is an interesting case in innovation research. First, the company is founded on the basic idea to turn cheese whey – three decades ago an invasive industrial waste problem – into useful production. Second, when the idea to produce whey proteins emerged some 30 years ago, it was a completely new business to enter. Hence, the technology, the production processes, the organization and the market were basically new. Third, the company was among the first two in the world to produce whey proteins on an industrial scale, and has maintained its leading position by continuously renewing its production base, thus becoming an important contributor to a growing, globally oriented ingredient industry. The idea of this chapter is that such a story can tell us something about innovation with care!

The original challenge in innovation research is associated with Schumpeter's engagement with understanding the emergence of new firms and their role in economic change. Schumpeter found that entrepreneurs possess the capability to be change agents in the economic system (Schumpeter, 1934). However, the classical entrepreneurial paradigm, and the subsequent techno-economic framework for innovation research, do not satisfactorily explain innovations in the modern economy. In short, a changing society also changes the nature of innovation, and the "innovation with care" approach is basically an argument that innovations involve all the complex and changing societal settings that any firm has within its field of operations.[1]

The line of thought followed here is that innovations are formed in interactive processes between a company and its surroundings (inside out and outside in scripts, see Sundbo and Fuglsang, 2006). Further, a firm may adopt different dominant modes of innovation due to different phases in its life course. Since Schumpeter's famous notion of "creative destruction", innovation studies have mainly been confined to the "change-element" in economic processes. But what keeps organizations unified, and what role does their internal structure play in innovation? This issue becomes even more important to investigate in a still more open system of innovation, one in which the resources of innovation become ever more diffused (Chesbrough, 2003). As care is a central element in innovative actions and strategies, it is relevant to ask what maintains continuity within change, and what this "caring element" means to the innovative capabilities of an enterprise?

The chapter is structured as follows: The business of whey protein and its Danish context is presented in the next section, followed by an analytical perspective on modes of innovation and the organizational balancing between continuity and change. Thereafter an empirical investigation of the development of Danmark Protein is outlined, subdivided into four main phases of its evolution.[2] The empirical findings are then discussed and analytically evaluated, followed by a brief conclusion.

WHEY PROTEINS AS A BUSINESS

Proteins are mammals' biological building blocks and the catalysts of all chemical reactions in living organisms. Since milk is the basic supplier of nutrients to the neonatal mammal, its nutritional and physiological importance is both vital and varied. Milk proteins have therefore been studied extensively for industrial purposes, and are probably the best characterized of all proteins (Mulvihill and Fox, 1994). Bovine milk contains 3.5 percent proteins, of which 80 percent is casein and 20 percent is whey. Casein has been used as a protein source for a longer period of time, while whey until the 1970s was only seen as a waste stream in the production of casein or cheese. Danmark Protein, one of the first two producers in the world, developed whey proteins from a pervasive industrial problem into valuable products, in three basic stages: whole-whey products, functional whey protein fractions and biologically active whey proteins. Their usage today covers a wide range of industrial areas (feed, food, pharmacy), because they possess many important functional, nutritional, physiological and pharmaceutical properties (Siso, 1996; Mulvihill and Fox, 1994). To mention a few examples, whey proteins may function as stabilizers and emulsifiers, or provide taste in a wide range of dairy, meat and sweet products. They may find use as specially designed food proteins, for instance, as substitutes for children at risk of allergic diseases, or as restitution ingredients for people involved in challenging sports activities, and so forth. These examples are only a few illustrations of the wide usage of whey proteins and their components in the industry.

Compared to other protein sources, whey protein is a difficult business, because it is a technologically difficult and economically more expensive substance to extract. Thus, it is extraordinary vital to find the functional advantages of the product and/or the niches in the market to exploit. As a consequence of these characteristics, whey proteins constitute a highly concentrated market segment, with only a few international players. Some 30 years ago, the initiators of Danmark Protein had only vague ideas about the functionalities and uses of whey proteins, and the technological and organizational difficulties they were facing. But Danmark Protein managed to survive a year-long process of exploration and setbacks to become the lead company within this very problematic business segment.

As argued in the introduction, the environment of organizational learning, understood as in-side out and as out-side in scripts, is vital to understand innovation with care. A simple explanation of this dual understanding of innovation activity is that the outside in script is about the ways in which an organization adapts to external condition, while the inside out script is about how an organization creates new ventures, and convinces others about the worth of a certain development (Sundbo and Fuglsang, 2006: 150ff).

Seen from the perspective of outside-in, the development of Danmark Protein cannot be separated from the traits of the Danish agrifood and biotech sectors. The evolution of Danmark Protein is only one example of a biotechnological business sector, that is the enzyme/protein sector, in which Denmark historically has created a strong position. Exploiting by-products from industry with biotechnological processing is an old practice in Denmark.[3] Several decades ago, the Christian Hansen and Novo Nordisk companies used calf stomachs and pig pancreas' as the raw materials to start up two of the most successful ventures in the biotech industry. Both of these companies have a long history as world leaders in the enzyme (protein) industry, due to their committed entrepreneurship, involvement, positioning and networking (Villadsen, 2005; Norus, 2004).[4] They have developed their leading international position through a long tradition of collaboration and competition with actors in different organizational and institutional domains, traits which are characteristic to the Danish enzyme business (Norus, 2004).

From the perspective of the inside-out script, it is interesting to note that Danmark Protein did not evolve within the core of the Danish biotech sector, which is in the Copenhagen region (Norus, 2004; Nelund and Norus,

2003). Instead, it developed in the heartland of the Danish co-operative movement in western Jutland. Even though the co-operative movement may share many of the same traits and characteristics of business practices as the bio-tech sector (Svendsen and Svendsen, 2000), it did not have any expertise in this kind of business, neither in its organization nor in its marketing. The fact that DP is a latecomer to the protein/enzymes sector, and had to start from scratch, makes its intra-organizational structure and management particularly interesting. How does one manage to continuously change by adapting and reorganizing, but still follow a technologically and organizationally new and difficult path? What kinds of structures, knowledge creation or even cultures have been important, and how do they affect the innovation process?

MODES OF INNOVATION AND THE BALANCING OF CHANGE

Innovation research has been dominated first by an entrepreneurial, then by a techno-economic paradigm (Sundbo, 1998). From the entrepreneurial perspective, innovations are personified in the individual who makes inventions profitable in a marketplace. The entrepreneurial paradigm is typically associated with the autonomous individual or "Gründer", who was so typical for firm development in the first industrial revolution, but they may as well be a community entrepreneur (Johannisson, 1987). In the techno-economic perspective, the main engine of economic change is technology itself. The reason for this has to do with the increasing role of, and the evolution of large scale Fordist production systems in the post war period. A main dispute in this paradigm was a discussion of whether "technology push" or "market pull" factors were the determinants of innovative change (Sundbo, 1998).

The significant changes in the world of production during the 1980s gave rise to a strategic imperative in business and innovation research (Sundbo, 1998), one in which the large companies normally were advised to concentrate on their core competencies (Porter, 1991; Prahalad and Hamel, 1990). Following Chesbrough's (2003) "open innovation" perspective, the role of the external environment became a more explicit concern in creating (internal) innovations. The general trend toward this perspective can be seen from the increasing outsourcing, vertical disintegration and networking activities of large companies. The trend toward new inventions in technology, new supporting structures and the "marketization" ideology are also important in this respect. Furthermore, the increasing role of services in the economy also helps explain the trend toward open innovation. For the single business, open innovation is a way to take advantage of the potentially global resources of knowledge, people and services, and internalize these to the advantage of one's own business. This trend has also regenerated entrepreneurship and small business development. These general trends are challenging the organization of innovation, because the internalization and maintenance of knowledge in an increasingly open system of innovation has become more problematic.

The analytical argument made here is that any organization, to survive, needs both stability and change. And both elements are important for their innovative capabilities. These two basic features of an economic organization can be contradictory, but they may also be complementary and synergetic. Change is vital for innovations, due to the strategic positioning and intentional behaviour of any firm in the contemporary economy. Strategies refer to the competitive element between companies, and are those intentional declarations by which a company signals its intentions to improve its development potential. Strategies may also derive from "irrational behaviour," like emotions and affections, or from mindful cooperation with other businesses in network structures based on reciprocity or association (Hovgaard, 2001). This "mindful cooperation" is often a hidden, stabilizing part of firm creation and development. I see it as the "coping" or "caring" element in innovations. But how can we analytically understand this?

Sundbo and Fuglsang (2006) and Fuglsang and Sundbo (2005) argue that the traditional understanding of innovations as either entrepreneurial or technology-science driven activities needs to be supplemented by a strategic-reflexive understanding of innovation. The intentional practices of strategy are not constructed from a simple calculative behaviour or logic, but are reflexively derived, that is from an understanding of a company's evolution as founded on an interactive interpretation of uncertainty, rather than objective calculation of facts (Sundbo and Fuglsang, 2006). "Strategy" is a rather broad term meant as a new conceptual tool to balance the change element in innovation, and to understand how external inputs are considered and maintained in a meaningful manner (Fuglsang, this volume). The interactive element, that is outside-in and inside-out scripts, is therefore expressed in the internal routines and habits and their interaction with external inputs and pressures, and is seen as particularly important in the modern "open innovation" business environment. Reflexive interpretation is immanent in any innovation, and is a way to generate order from chaos, in the form of data, information, or knowledge, to create strategies and action. The challenge here is to understand how the typical "chaos in change and innovative processes" is balanced by elements of "stability" as for instance norms and routines for organizational behaviour.

The need for balancing stability and change can be interpreted with March's (1991) distinction between a company's need to balance the essential features of exploitation and exploration. Exploration refers to things like search, risk-taking, flexibility and innovation (March, 1991: 71). Its essence is "experimentation with new alternatives. Its returns are uncertain, distant, and often negative" (March, 1991: 85). Exploitation refers to things like refinement, choice, production, efficiency and implementation (March, 1991: 71). Its essence is "the refinement and extension of existing competencies, technologies, and paradigms. Its returns are positive, proximate and predictable" (March, 1991: 85). There are no fixed ways in which a company can balance these two; the balance will depend on the organizational context, in the form of the pressure of the external environment and how norms, routines and practices within the organization are reflexively derived and structured.

A way to conceptualize how routines and norms become reflexive practices can be extracted from Thompson's (1996) framework of tradition. Thompson distinguishes among four modes of tradition. The first one is normative tradition, which is a form of tradition in which assumptions, beliefs, routines and practices from the past are taken for granted. Normative tradition is about "making things as we always have," and can be explicit and strong, meaning that the past is explicitly referred to, and should not be questioned, as in Weber's traditional authority. Or normative traditions may be weak, in the sense that they are un-reflected practices, as for example "we do things this way, but we do not know why we do them so." The second mode of tradition is legitimate tradition, and is about Weber's scheme of legitimacy which distinguishes between beliefs in legitimate power (Weber's rational legitimacy), like regulatory bodies. Or it is belief in something exceptional (Weber's charismatic legitimacy), like a distinguished entrepreneur. These first forms of tradition have been the dominant ones during history, but they are now declining and being replaced by the reflexive traditions of interpretation and identity making. In the interpretative (or hermeneutic) form of tradition, the taken-for-granted assumptions in one's life function as a scheme for constantly understanding or interpreting the world. In identity making, the past is a source of personal identity creation, either as self-identity or as an individual's sense of being part of a community, that is collective identity.

THE EVOLUTION OF DANMARK PROTEIN

As any other kind of industrial production, agriculture has a wealth of different by-products to deal with. Whey is a milk white water from cheese production, which counts for 90 percent of raw material input (milk). In other words, from 100kg of milk, you get 10kg of cheese and 90kg of whey. The structural changes in the agrifood business in the late 1960s and 1970s put more emphasis on value added products like cheese, which explains the large increases in quantities of whey by-product. The whey was normally returned to the farmer as a fodder supplement, that is as a low value product, but the increasing quantities of whey became much greater than could be absorbed. Furthermore, longer distances were involved, with the closure of many local dairies. This was true for Denmark, as well as other European countries (Mulvihill and Fox, 1994: 9). For these reasons, much of the residue of cheese production had to be left on the fields, simply as waste. This practice quickly became an environmental problem. The fact that whey had an interesting biochemical composition had long been recognized, including the possibility to use it as a source for human consumption (Siso, 1996; Mulvihill and Fox, 1994; Whittier, 1948).

The Danish (and European) dairy industry changed dramatically in the post-war period. In particular, the late 1960s and the formation of the European Community enforced great organizational changes, that is concentration and economies of scale (Bjørn, 2001). These institutional changes also affected the decision of a few local dairies in the south-western part of Denmark to establish a joint company "MD-Foods" in 1970. MD-Foods quickly gained a strong position on the Danish market, and after several years of "milk-wars," it became the single strongest dairy company in Denmark (Søgaard, 1990). MD-Foods, like many other dairies, had problems getting rid of their whey, and the growing quantities of the residue and its environmental consequences made them interested in using the whey for industrial purposes. They knew that it was possible to dry the whey for feed purposes, but also that this process was an unprofitable venture.

The Pioneer Phase

The first realistic idea to industrialize whey came by coincidence. Two Swedes and a Dane visited MD-Foods head office in 1973, as advised by the dairies' organization (Danske Mejeriers Fællesorganisation). One of the Swedes had a patented technical solution to extract whey proteins, but had no access to realistic quantities of the product.[5] The management of MD-Foods found the idea exciting, and by personal contacts, they found a pharmaceutical company interested in joining the project. From MD-Foods a few key people joined the project, first and foremost being Jørgen Krabbe, who was the Head of Planning in MD-Foods at the time. The new partners started to cooperate in late 1973. By the spring of 1974, they formed the company Novitas Danmark A/S, but the name was legally challenged by a third party, and subsequently changed into Danmark Protein A/S in 1975.

A defunct dairy outside the village of Hobro was used as an experimental station, and a pilot plant was constructed in which the first tests and experimental production could be done. The researchers managed to extract a highly purified protein (over 90 percent), but it was a troublesome and expensive production process, particularly the freeze-drying. The personal contact from the pharmaceutical partner had left, and the new pharmaceutical management showed no interest in investigating the product. Although they had an excellent product, they saw no market, and did not have the means or skills for a seven-year-long development process, which was seen as the time span necessary to enter the pharmaceutical market. The problems divided the partners over directions in which to move. Furthermore, the working capital was exhausted. However, with Mr Krabbe as the prime mover, they began to explore new directions, and the result was that MD-Food took over all shares in the company.

Although the first period of exploration was not a successful one, it provided knowledge for further initiatives. It was realized that another technological track and another market was necessary. By chance, one of the engineers at MD-foods knew that the sugar company De Danske Sukkerfabrikker (today Danisco) was developing its own membrane technology aimed at improving its own sugar production. But sugar has a high tendency to crystallize, and therefore the membranes at that time did not work very well in extracting sugar. With milk it was a bit different, and membranes in particular proved to be efficient in extracting water from milk. Actually, the introduction of membrane technology on an industrial scale was the precondition for the so-called "Feta-Adventure," that is the large export of feta cheese from Denmark to eastern countries during the 1970s, particularly Iran.

Membrane technology became the strategic technological key for a new direction, and this made it possible to replace freeze-drying by spray-drying, which is a considerably less expensive drying method. Production facilities were established in the basement at the condensation factory HOCO in the city of Holstebro, and many experiments had to be carried out with this new technology during the summer of 1976. But they succeeded in starting small-scale production with a purity maxima of 80 percent.[6] Having relative control over the technology process, DP was ready to search for a market. Krabbe had hired a sales agent, and together they managed to find a Japanese purchaser of high quality protein extract (over 75 percent in purity). Selling high quality proteins was important, because their relatively high production costs demanded higher prices. The purchaser was the industrial giant Mitsubishi, and when the first order finally came, it was for the significant amount of 20 tonnes. The problem was that DP failed to produce enough quantity with the purity demanded, and within the agreed timeframe. In addition to struggling to meet production demand, they had to use all their skills and capabilities to explain the situation to the purchaser. DP succeeded in delivering this first big order, and Mitsubishi became their main purchaser of high quality whey proteins for several years.

Having control of the technology and a growing market, they expanded the membrane system as far as the space in the basement allowed.[7] But the idea of constructing a stand-alone plant had started to mature. To reach this goal, there were other problems. First and foremost, they realized that commercially viable production had to generate rather high quantities. The main problem here was the huge permeate residue, which accounted for 80 percent of production outcome, but had no market, nor a realistic way to dispose of it. Also, all parties knew it was still a risky business, and therefore there was no easy solution to find the large investments needed. Again, a long process of exploration was needed to find technical and organizational solutions.

Again, one of the main problems was resolved accidentally. At a trade exhibition in Germany, Mr Krabbe met a salesman who showed great interest in this new protein business. He represented Kali Chemie AG, a German chemical-pharmaceutical company looking for new investment opportunities; they even found the permeate particularly interesting, planning to use it as a fire retardant in their production of insulation materials. These individuals soon realized that there might be synergetic effects available through joint venture, and an immediate personal friendship between Mr Krabbe and the head of the biotech division for Kali Chemie helped to strengthen the connection.

On the basis of this new partnership, an ambitious plan to construct the first real whey protein plant in the world was adopted in 1978. Kali-Chemie was willing to hold 50 percent of the shares. Furthermore, they invested in a separate factory close to Danmark Protein to produce the permeate into lactose.[8] MD-Foods management, which had supported the project all the way through, did not see themselves able to invest the remaining 50 percent by itself. Therefore, Mr Krabbe had to use all his network capabilities to find other investors; he managed to have the dairies organization and three dairies join in.

When the basic organization was in place, things moved rather quickly, and a new plant with the capacity of 1000 tonnes a day was ready to operate in the second quarter of 1980.[9] The plant was placed in the town of Nørre Vium, and its exciting undertaking quickly attracted skilled workers from the whole region.

Large Scale Producer

The modern, new plant faced many start up problems. It was quickly realized that there was a bacteriological problem, since the dairies that delivered the raw material were used to treating the whey as waste. Numerous resources were therefore put into quality treatment at the dairies. This was no easy process, because the dairies had to make bigger investments to reach quality demands. Second, the DP-Kali complex was bound by agreement to accept the whey. Krabbe bought a condensation factory just to dispose of the poor quality whey.

The second main problem was in processing itself. The plant was working well the first few weeks, but then production interruptions occurred regularly. The material suddenly had a tendency to seize during operation, which demanded numerous resources in cleaning and repair. This was the first factory of its kind in the world, so a highly interactive process between technology and processing experts was needed. Again, it was possible to attract people with high competencies to deal with processing and techniques. Also, the German partner had great expertise in industrial production, and their willingness to share this knowledge proved to be invaluable. The main problem was a hidden one, in the sense that no one could tell whether the interruptions were seasonal or technical. Eventually, some of the employees figured out that the interruptions happened a few weeks after the farmers changed fodder for the cows. Having found the source of the problem, they were able to adapt the system. Furthermore, other processing elements, such as temperatures and air moisture were adjusted.

The adaptation of the technical problems mentioned lasted for more than three years, and the protein generally reached a higher level of quality so that the condensation factory could be sold off again. The adaptations exemplify a change toward exploitation rather than exploration in the company's development. Another important element in this is that there was a need to construct whole new systems of control and management: budgets, finances, settling of accounts, and the like. Furthermore, as the basic management tools were put in place, strategic planning was introduced as a new tool to continuously analyse the situation and direct the development of the company. But even on these issues, Kali-Chemie had an expertise they were willing to share.

Due to the many start up problems, the company had to face big deficits during the first years of production, and the shareholders were required to cover the deficits. Again, the German partnership proved invaluable, because they had the attitude that deficits and operational problems were an integral part of the game. The local dairies had the opposite view, because they were used to operating technology that worked from day one. An intermediate, but supportive position was held by MD-Foods, which had their own experiences in introducing new management styles in the Danish dairy business. To maintain the unity of the company, the management tools were important, since they provided the necessary bonds among the parties.

With the elimination of technical problems, the quality of the protein also increased, and so did demand. The first profitable year was 1983. During that year, they increased production capacity to 300 000 tonnes of whey a year. In 1984, Mr Krabbe realized that he had succeeded in seeing his endeavours grow into a full-fledged enterprise. He therefore decided to leave, and handed over control to the financial manager Thorkild Stotz.

In 1986 they increased production capacity to 600 000 tonnes of whey, where the whey protein isolate was at the 90 percent + level; the first trials to develop peptides by enzymatic hydrolysis were now undertaken. In other words, the first steps to diversify and improve their production base were taken. One reason behind these new exploratory steps was based on the fact that other actors had entered the market, but also on the fact that they had known for years they were too dependent on one customer. This dependency turned into a considerable problem when the Chernobyl catastrophe occurred in 1986. Japan heavily increased their standards for food imports from Europe, and in practice – over night – created an import ban on all milk proteins. On the other hand, the new Japanese standards helped trigger a new phase of exploration.

Innovation as Principle

A set of interrelated strategic changes took place at Danmark Protein in the latter part of the 1980s, ones which promoted a new phase of innovativeness. First, the organization was changed so that production and management, which until then had been in different places, were merged at the production facilities in the village of Nørre Vium. The team that was already emerging under Mr Krabbe's leadership was more clearly organized within one physical structure. This merger also helped in creating a distinctive "DP-spirit"; one which was important for new ideas to flourish.

Second, there were new markets for specially designed proteins, particularly in Japan. This shift also demonstrated the importance of having experienced people in new areas. A manager for the new sales department was selected for his international experience, including his suggestion for an open dialogue with the Japanese market. Practically, this meant that products should be developed in concrete dialogue with the customer, and aimed at their specific needs. The Japanese market was very demanding, both technologically and culturally. The Japanese had impressive knowledge of the theoretical possibilities for whey proteins, and they were world leaders in developing meat and dressing products by using egg and soy proteins. To match this challenge organizationally, a DP sales office was opened in Japan.

Third, a focused R&D strategy was started to take further advantage of the functional and nutritional characteristics of the whey protein. This strategy had a modest start with a one-person research department in 1987. An applications department was also established, which had a more practical objective, that is the task of mixing and testing products before they entered the market. This department proved to be imperative in supporting the sales organization, and as a testing ground for new inventions. Furthermore, development in membrane technology made it more flexible, and therefore created more possibilities for experimentation. For example, a membrane pilot plant survived a failed development project with another Danish company, but could be used for many other purposes of experimentation by the flexibility of this technology.

Fourth, a quality assurance system was implemented. Even in the mid-1980s it was apparent that there were too many differences in product outcome, although people thought they were doing precisely the same kind of processing. For some time the management worked on its own quality system, but it was problematic, since there was no comparable production or production system with which to compare it. After the introduction of the ISO-system in the late 1980s, management decided to certify the company according to this standard. While this certification process was a prolonged and demanding task, by 1992 the company was approved under the ISO 9002 standard, and in 1994 with the 9001 standard. By 1998, the company was certified according to the Danish Standard 3027 (the HACCP standard).

In this period we can observe a dynamic balance between exploitation and exploration. This is a period in which the company constructed structures to maintain and develop its own internal learning capacity and knowledge. This is another reason why it is important to have a united organization, as development and innovation involves cooperation among development, production, application and sales. This is very much based on inside out scripts. Some of the inventions are made by internal experimentation, while the inspiration for many others came from people working in the market, that is the sales department. There is a dialogue with the market, but development comes from within the organization. Top management is another "stabilizing" element, as they had to legitimize high budgets for product development. Strategy is continuously changed and adapted, an exciting but demanding process which also helped in creating stability to the organization.

Importance

The strategic shift and the subsequent reorganization in the late 1980s, combined with a maturing protein market, helped Danmark Protein develop quickly during the 1990s.

Another sales organization was established in Germany in 1990 (Danmark Protein GmbH), and a third in North America in 1991 (Royal Protein Inc.). In 1991, the market capacity increased to 1,2 million tonnes a year. By the end of 1993, DP bought up 50 percent of the shares in the German company, Biolac GmbH. This particular move gave them the means to enter a new market, and to anticipate that in Denmark there would soon be a short supply of whey. Also, in the late 1990s, technology was adapted to the still more advanced forms of production, and the whole membrane system was renewed.

Despite the fact that DP was a growing business at the beginning of the 1990s, interest from the German partner diminished. Kali was restructuring its organization, and following the business ideas of those days they wanted to stick to their core competencies. At this point in time, Danmark Protein had a strong and innovative organization, but it is indisputable that without the knowledge and competencies that Kali brought to Danmark Protein, the company would probably never have succeeded.

At that same time the exhausting "milk-wars" between the two big companies within the Danish dairy branch – MD-Foods and Kløver Mælk – enforced cooperation. To buy out the German partner, a common company created from these two antagonists acquired all the shares in DP in 1994 (Midtjyllands Avis, 19 July 1994). In the accounts the year before, that is 1993, DP had a turnover at 320 million DKK and 220 employees. By comparison, the turnover during the first year of production was about 8 million DKK (Ringkjøbing Amts Dagblad, 30 December 1993).

Global Developer

From the late 1980s to the early 1990s, about one new product category was established each year. But from the early 1990s, product development takes off, and around the year 2000 Danmark Protein had about 200 different product categories. There were more than 40 employees working on product development and applications; the two units were, in practice, turned into an "Innovation Center."

As already argued, the market for proteins changed a great deal in this period, and so did the dairy business as a whole. In Denmark, MD became the sole winner of the Danish "milk wars" and acquired most of its competitors, Kløver Mælk included. But still another development was increasingly felt, which was the increasing globalization of the dairy business. The retail stores at the user-end of the agricultural commodity chain were getting bigger, stronger and globally integrated. They made major demands on the producer end of the chain, and one of these demands was to deliver a broad range of product categories. To meet these demands MD-Foods organized a new ingredients department; one which resulted in the functional integration of DP into a bigger organizational structure. This was required by the need to put more focus on ingredients and the increasing need for customer focus (Ringkjøbing Amts Dagblad, 7 January 1997). In practice, this meant that DP's days as an "independent company" were over, an organizational change that many employeees found problematic. Another consequence of the dairy business "going global" is that MD Foods and the Swedish dairy giant Arla in 2000 were merged into Arla-Foods in 2000. Arla-Foods became one of the biggest dairy companies in Europe.

Today, DP is a production unit within the Arla Global Ingredients division, which manages all dried products for the company, and is integrated with other divisions at Arla headquarters in Aarhus. But product development still needs to be close to production, and the innovation center at Danmark Protein is still there. The strengths that created the success of Danmark Protein from the mid-1980s to the mid-1990s have disappeared, but new times demand new forms of organization. Being part of a global supplier of dairy products, the whey protein unit is crucial to develop new products and to establish the nutritional and functional values of the proteins.

Even more protein products are developed in direct relations with the customers, and these involve still broader networks. Today, Arla has at its disposal a range of specialized development centers, which together form an inter-organizational network of innovation, which either work on internal ventures, or in an open system of innovation with external business or institutional partners. One example of this internal networking is the recent development of the Arla Mini30 milk, which is the first milk in which components other than fat were changed. The milk contains no milk sugar, but still contains proteins, calcium and other vital components. Another example is the development of Lactolite, a lactosis free milk for people otherwise intolerant to milk.

One example of moves toward external and open systems of innovation is a researcher at Aarhus University detecting that a protein (OPN), first discovered in the mid-1980s, was also contained in milk. Close co-operation among the parties involved has resulted in three new patents. Another example is the recently started nanofood consortium (www.inano.dk/nanofood), where Arla (including DP) is only one partner of many in researching and developing the possibilities of nanotechnology, which is seen as a new path toward developing the agrofood business.

CAREFUL INNOVATION BY MINDFUL ORGANIZATION

The first ten years of DP's history are characterized by the entrepreneurial mode of innovation, practiced through processes of trial and error. First and foremost the charismatic leadership of the manager, Jørgen Krabbe, led the strategic positioning. From the very beginning, he was strongly committed to the ideas about whey. In addition, he had an impressive ability to attract competent, stable people to the organization. Furthermore, he is the one who managed the two vital international links, that is the Japanese market and the collaboration with Kali-Chemie.

But business ideas are also nurtured within a specific environment. At the beginning of the 1970s, many organizational and technical experiments were taking place in the Danish dairy industry, and market conditions were forbidding. MD-foods was among the first to have professional business units to deal with planning, market analysis and other management systems, which certainly explains some of its success over most other national competitors. These circumstances, together with changes in the production structures of the dairy industry nurtured an environment in which a company like Danmark Protein could emerge. Furthermore, the expansive strategy of MD-Foods made them "the naughty boy" for years in the Danish dairy business. This drive to be on one's own in a tough environment provided buffers for errors in Danmark Protein, as well as support for moving in new directions. Also, the fact that DP was not directly thought of as a profit-making engine, but rather as a necessary precondition for the success of the dairy industry also diminished expectations, and provided necessary tolerance. The combination of German industrial experience, the cooperative network and the new professional management style at MD-Foods are the general threads that created a strong DP organization. MD-Foods and Kali-Chemie functioned as intermediaries through their financial and institutional support, without which DP would never have survived the early, entrepreneurial years.

The early years of exploration and entrepreneurship were replaced by a shift toward exploitation and organizational learning, which began around 1980 (that is techno-economic or institutional mode of innovation). Typically, exploration and exploitation are seen as more or less unrelated (March, 1991). However, Meeus *et al.* (2006: 133) argue that exploratory activities feed the exploitative part of organizational learning, a fact which can also be observed in the first phase in the evolution of Danmark Protein. It is very clear that the exploitative practices from the early 1980s, in the form of different management structures, created the necessary stability and security for the organization at a point of time it was particularly needed. DP then begins a technological trajectory, and there was a need to institutionalize routines and standardize goods. Though, this period should not be seen as one of techno-economic adaptation only. It is also the interpretation and generation of an independent DP culture, one which helps to make the different divisions cooperate on the basis of common values and strategies.

Compared to the earlier entrepreneurial mode, innovative and explorative activities are internalized, and take on the shape of techno-economic innovations in push-pull relations with the market.

The techno-economic mode of innovation reaches its limits in the early 1990s. Despite this DP was capable of controlling and advancing its products, even given an unpredictable economic future in the context of a new mode of open innovation. Furthermore, the more profound the internal knowledge and competencies of whey proteins become, the more the complexity of new challenges promote a system of open innovation. Integration into a bigger ingredients division is an organizational way forward, and another procedure is to put even more emphasis on strategy and reflected interaction in networks, as exemplified by the organizational division within the Arla concern, and Arla's connections with national and international institutional bodies.

From the mid-1990s, DP was no longer an "independent" company, although its features are still visible in the form of the learning structures and traditions it created. DP's influence is also manifested in the continuation of the innovation center at the production site. But development and innovation are now part of a complex organizational structure in which the different units both have to cooperate and compete to receive necessary funding. This makes interaction and reflexive interpretation on all levels of management a continuous challenge to maintain stability within change.

CONCLUSIONS

Today, Arla is one of the big players in the global dairy market. Its branch Danmark Protein is only one, crucial, node in a dense organizational network, which provides vital competences and know-how. But without the links back to a basement in Hobro, Arla's leading position within this business segment would have been substantially smaller. Danmark Protein was the only really "true innovation" within the Danish dairy sector for decades, meaning that the organization, the technology and the products were basically new. Despite this challenge, it matured into an excellent internationally well reputiated and profit making enterprise. The idea of this chapter was to exemplify innovation with care from the perspective of this extraordinary story, and its main achievements can be summarized as follows:

- Innovation is an activity that changes between different business modes, primarily due to changes in the external environment and the ways in which external relations are subsumed in the organizational structure and culture.
- Exploration and exploitation are activities which may vary, due to different innovation modes. Finding a proper balance between exploration and exploitation in the changes from one technological track to another is key in the success of a company.
- The implementation of an interpretative tradition (that is reflexive tradition) of mutual networking and common understanding is the practice that creates a balance between exploration and exploitation.

NOTES

1. As a first-stage theoretical proposition to understand innovation in this broad sense, the case study approach is particularly valuable, since it is a scientific tool aimed at an in-depth investigation of one single phenomenon or question. Case studies are sensitive to, and useful for, the testing of theoretical propositions and/or the development of new theories (Hovgaard, 2001; Yin, 1994; Glaser and Strauss, 1967).
2. The empirical findings of the chapter are based upon interviews with leading executives in the development of the company Danmark Protein, on archival materials kindly given to me by the management, and on newspaper articles and other materials found in Danish library archives.
3. Byproducts or waste are the remaining materials from one kind of industrial production, which, at best, only find low value usage in other industries. Most of these remaining are disposed of, dumped (typically into the sea), or "disappear" into the air (smoke). The processing of these byproducts frequently cause severe environmental problems.
4. In contrast to other fields of biotechnology, Europe actually does better than the US in applying micro-organisms in the food sector (Valentin and Jensen, 2004: 167).
5. One problem in extracting proteins is that they cannot tolerate too much heat-up, and therefore have to be extracted by means other than the ones usually known within dairy production.
6. The more purity of the protein, the better segment in the market, human consumption for instance.
7. Actually, ultra-filtration became the technological precondition for a commercial breakthrough of whey protein production (Mulvihill and Fox, 1994: 9f).
8. The idea of using the permeate as fire retardant was never realized, but other opportunities were found.
9. The initial investment turned out to be 75 million DKK. This was considerably more than originally planned, but Kali-Chemie realized that it would be too expensive to transport the permeate to Germany. Thus, they constructed a plant of their own at the same site as Danmark Protein. And, indeed, the two companies were operated as one.

REFERENCES

Bjørn, C. (2001), *100 år for dansk mejeribrug*, De danske Mejeriers Fællesindkøb.
Chesbrough, H.W. (2003), *Open Innovation*, Boston, MA: Harvard Business School Press.
Fuglsang, L. and J. Sundbo (2005), "The organizational innovation system: three modes," *Journal of Change Management*, **5** (3), 329–44.
Glaser, B.G. and A.L. Strauss (1967), *The Discovery of Grounded Theory: Strategies for Qualitative Research*, Chicago, IL: Aldine.
Hovgaard, G. (2001), "Globalisation, embeddedness and local coping strategies. A comparative and qualitative study of local dynamics in contemporary social change," Ph.D. thesis for Roskilde University.
Johannisson, B. (1987), "Anarchists and organisers. Entrpreneurs in a network perspective," *International Studies of Management and Organisation*, **17** (1), 49–63.
March, J.G. (1991), "Exploration and exploitation," *Organizational Learning*, **2** (1), 71–87.
Meeus, M.T.H., J. Faber and L.A. Oerlemans (2006), "Do network structures follow innovation strategy?" in J. Sundbo, A. Gallina, G. Serin and J. Davis (eds), *Contemporary Management of Innovation*, New York: PalgraveMacmillan, pp. 131–46.
Mulvihill, D.M. and P.F. Fox (1994), "Developments in the production of milk proteins," in B.J.F. Hudson, *New and Developing Sources of Food Proteins*, London: Chapman & Hall, pp. 1–30.
Nelund, R. and J. Norus (2003), "Competencies and opportunities – building an island of innovation apart from Europe's innovative centre," in H. Ulrich (ed.), *Regionalisation and Globalised Innovation*, London: Routledge.
Norus, J. (2004), "Building regional competencies – the industrial enzymes industry," Copenhagen Business School Research Centre on Biotech Business working paper no. 07.
Porter, M.E. (1991), "Towards a dynamic theory of strategy," *Strategic Management Journal*, **12** (special issue, Winter), 95–117.
Prahalad, C.K. and G. Hamel (1990), "The core competence of the corporation," *Harvard Business Review*, **68** (3), 79–91.
Schumpeter, J.A. (1934), *The Theory of Economic Development*, Cambridge, MA: Harvard University Press.
Siso, M.I.G. (1996), "The biotechnological utilization of cheese whey: a review," *Bioresource Technology*, **57** (1), 1–11.
Søgaard, V. (1990), *Spildt mælk? – en analyse af konkurrence, samarbejde og solidaritet mellem landbrugets andelsvirksomheder*, Esbjerg: Sydjysk Universitetsforlag.
Sundbo, J. (1998), *The Theory of Innovation: Entrepreneurs, Technology and Strategy*, Cheltenham, UK and Lyme, US: Edward Elgar.
Sundbo, J. and L. Fuglsang (2006), "Strategic reflexivity as a framework for understanding development in modern firms," in J. Sundbo, A. Gallina, G. Serin and J. Davis (eds), *Contemporary Management of Innovation*, New York: Palgrave Macmillan, pp. 147–66.
Svendsen, G.H.L. and G.T. Svendsen (2000), "Measuring social capital: the Danish co-operative dairy movement," *Sociologia Ruralis*, **40** (1), 72–86.
Thompson, J.B. (1996), "Tradition and self in a mediated world," in P. Heelas, S. Lash and P. Morris (eds), *Detraditionalization*, Cambridge, MA and Oxford: Blackwell Publishers, pp. 90–108.

Valentin, F. and R. Jensen (2004), "Networks and technology systems in science-driven fields: agriculture," in M. McKelvey, A. Rickne and J. Laage-Hellman, J. (eds), *The Economic Dynamics of Modern Biotechnology*, Cheltenham, UK and Northampton, MA, USA: Edward Elgar, pp. 167–206.

Villadsen, J. (2005), Dansk Bioteknologis historie – de første 75 år. BioZoom, 4, Biokemisk forening, accessed at www.biokemi.org/biozoom.

Whittier, E.O. (1948), "Mejeribrugets Biprodukter," *Nordisk Mejeri-Tidsskrift*, **14** (6).

Yin, R.K. (1994), *Case Study Research: Design and Methods*, Thousand Oaks, CA: Sage.

7. Public innovation with care: a quantitative approach

Lars Fuglsang, Jeppe Højland and John Storm Pedersen

INTRODUCTION

Innovation with care is an approach to innovation that investigates some of the tensions that exist in innovation processes between creativity and innovation (Amabile *et al.*, 1996), exploration and exploitation (March, 1991) or, as we will emphasize in this chapter, variation and selection (see Nelson and Winter, 1977).

According to the model discussed here, capturing benefits from innovation requires, on the one hand, that a variety of creative approaches and ideas exist inside a particular domain of development and innovation. On the other hand, it also requires recognition, selection and aggregation mechanisms that ensures that some of those ideas are selected, assuming that some of them are better or more appropriate than others.

We can distinguish between a domain of innovation such as an institution (like a public school) in which creative ideas, variation and selection may exist, and a field of innovation in the broader context of that single institution, where recognition and selection of important ideas takes place (see Csikszentmihalyi, 1996 see also the chapters by Møller and Fuglsang in this book).

Recognition and selection which takes place in the broader context of innovation is a mechanism that can be understood both as a driver of innovation and as a mechanism for capturing the benefits of innovation. It is a driver of innovation because it pays attention to and recognizes distinctive innovative activities, for example by rewarding innovative people. And it implies selection, diffusion and exploitation of some of the ideas while others are dismissed as less important. We consider selection mechanisms of this kind, at the level of the field, to be an important characteristic of the innovation process that can take different forms depending on the specific context, for example in the public sector or in the private sector.

Innovation with care is also an approach that assumes that innovation is an interactive process that involves many and changing actors in an exchange of opinions and ideas. This requires that actors and perspectives are carefully woven together. The reason for this is that interaction takes time and that many people and their opinions and experiences have to be adjusted to one another. This corresponds to Popper's notion of piecemeal change as opposed to utopian social engineering (Popper, 1962). The implication of this is that the selection process is often stepwise, inspirational and heterogeneous.

As for the public sector and public innovation, it could be argued that care, variation, creativity and selection at the level of the domain are already mechanisms inherent to the public sector: the public sector consists of many locally anchored institutions and professionals with a high degree of autonomy to act on their own.

The more difficult point for the public sector is perhaps selection at the level of the field: to scale up and select the important ideas across institutions. What selection mechanisms exist here? Indeed, we may argue from simple observations of institutions that – as a matter of fact – certain field mechanisms of selection do exist, because institutions often adopt the same types of idea. Such mechanisms do not have to be mechanisms of homogenization and isomorphism, which is sometimes implied by institutional theory (DiMaggio and Powell, 1983; Meyer and Rowan, 1977. See also Fuglsang in this volume). They can also represent attempts to adopt the more important, relevant and appropriate ideas. Furthermore, this kind of selection does not necessarily mean that institutions adopt exactly the same ideas or standards: Institutions can be inspired by each other and they may learn from each other and the good examples through certain mechanisms of selection.

We could tentatively formulate this seemingly paradoxical phenomenon in the following way: the selection mechanisms are not always "convergent mechanisms," leading institutions to adopt exactly the same ideas, but sometimes divergent, which means that institutions learn from each other (or learn from the good ideas and practices). Still, the selection mechanisms of divergent selection may sometimes be very difficult to establish and elaborate in practice in the public sector.

In this chapter, we begin by presenting some results from a survey regarding innovation in public institutions in Denmark (Pedersen, 2007). We then define a proxy for innovation with care at the level of the individual organization (the domain) and investigate the impact of innovation with care at the level of the field. The result of this analysis can be summarized in the following four points:

1. the leaders of the public institutions in Denmark report that public institutions innovate;

2. innovation seems based, to a large extent, on the involvement of employees;
3. variations between institutions are the rule since very little copying appears to go on; and
4. institutions which adopt an approach of "innovation with care" seem better at satisfying external requirements at the level of the field.

In addition to this, it appears as if institutions, which adopt an approach of innovation with care, also have a special relation to the political-administrative system. These points will be elaborated below.

After this, we discuss how the selection mechanisms in the public sector can be characterized. If we are right, selection can be characterized by divergent rather than convergent selection across institutions. First, we find that some mechanisms of convergent selection actually exist at the level of the field based for example in consumer choice or minimum standards. Second, we outline, at the theoretical level, various examples of divergent selection such as strategic arenas, institutions for professionals and media attention.

PRESENTATION OF SURVEY

The survey includes responses from 759 leaders of Danish public institutions – kindergartens, schools, after-school recreation centers, special institutions and homes for the elderly. These institutions make up the core of the production and delivery of services in the Danish public sector. The survey was conducted in November 2006 and includes questions regarding success criteria, governance structures, dialogue with the political-administrative system, relations with employees and innovation.

Stratified cluster sampling was used. The clusters (98 municipalities) were stratified in terms of geography (five regions) to ensure representation from all parts of the country. In each region six municipalities were selected based on "probabilities proportionate to size" (PPS) to ensure proper representation of institutions in large municipalities (De Vaus, 2002: 75–6). Finally, 50 institutions were randomly chosen in each municipality.

The PPS method means that a municipality with 400 institutions has a four times greater chance of being selected than a municipality with 100 institutions. Thus the large municipalities have a higher probability of being chosen, but since the same number of institutions in each municipality is chosen regardless of size, this is compensated for by the lower probability of a particular institution in the large municipalities being chosen.

Data collection took place via an Internet survey. The leader of each institution received an email including a hyperlink leading to a web page where the survey could be completed. There was a 51 percent response rate which is an acceptable rate making it possible to generalize the results to the population of around 11 000 institutions.

PRESENTATION OF RESULTS

More than two out of three leaders reply that their institution has innovated within the last five years (see Table 7.1).

Most of the innovating institutions (70 percent) state that innovation was self-initiated, 23 percent report that the new production and delivery of services was developed in collaboration with other institutions, whereas only 7 percent of the institutions state that innovation was based on copying or intervention from third parties.

Another finding of the survey is that employees play a major role in innovation in public institutions. Out of 446 innovating institutions, 78 percent report that employees "to a great extent" were important for the development of new ways of producing and delivering services. This result becomes particularly remarkable when comparing it with the responses regarding the role of other actors. Professionals and the political-administrative system are reported to be important for the innovation process "to a great extent" by 20 percent of the institutions. Consultants, ministries, other public institutions, and universities are of even less importance to the innovation process.

To sum up, two-thirds of the public institutions report that they innovate. Innovation is mostly self-initiated, and employees are by far the most important resource in the innovation process. In other words, innovation in public institutions is rather widespread and is for the most part internally driven, and variation across institutions rather than copying between them is the rule.

In order to analyse some of the differences among the innovating institutions, we define a proxy for innovation with care at the level of the single

Table 7.1   Innovation in public institutions

Question: Have there been major changes in the way your institution produces and delivers services within the last five years?

	Percentage	N
Yes	68	446
No	32	210
Total	100	656

institution and look into the effects of this innovation with care on different output measures. To be included in the group of institutions innovating with care, the leaders/institutions must fulfill five criteria. This is to say that inclusion in the group is based on five items from the questionnaire (see Table 7.2).

Out of 446 innovating public institutions, 146 or 33 percent of the institutions fulfill all five criteria and are thus defined to innovate with care. The five criteria for inclusion focus on two main dimensions: the involvement of employees in the innovation process and the criteria of the leader for selecting ideas.

The first criterion eliminates the institutions where the relationship between the leader and employees are within rather fixed limits – that is the traditional industrial model. The reason is that the latter kind of relationship inhibits the inclusion of employees, crucial for innovation with care. The second criterion eliminates the institutions where employees were not involved "to a great extent" in the innovation process. The third criterion focuses on the involvement of users. The fourth and fifth criteria eliminate the institutions with a leader who does not "to a great extent" find it important to base the production and delivery of services on professional criteria and on transparency.

Table 7.2   The proxy of innovation with care

Question	Innovation with care	Others
1. What relationship do you have with your employees regarding the production and delivery of services?	An open exchange/A partnership	A stabile exchange
2. To what extent were your employees important when new ways to produce and deliver services were developed?	To a great extent	To a certain extent/To a low extent/Not at all
3. To what extent is it important that your institution produces and delivers the services in dialogue with the users with the aim to deliver tailored services?	To a great extent	To a certain extent/To a low extent/Not at all
4. To what extent is it important that your institution produces and delivers services on the basis of professional criteria and not on the basis of economic criteria?	To a great extent	To a certain extent/To a low extent/Not at all
5. To what extent is it important that your institution has transparent criteria for the production and delivery of services?	To a great extent	To a certain extent/To a low extent/Not at all

Table 7.3 Innovation and fulfillment of external demands

Public institutions where innovation has led to a better fulfillment of external demands, by type of innovation. Percentages and estimated odds ratios. Danish public institutions in 2006.

	Better fulfillment of external demands	
	Percentage	Odds Ratio
Total	49	
Type of innovation		***
Innovation with care	62	2.5
Others	42	1.0
N	402	

Source: Danish public institutions. November 2006.
Significance levels: *: $(0.01 < p \leq 0.05)$, **: $(0.001 < p \leq 0.01)$, ***: $(p \leq 0.001)$.

The institutions which fulfil these five criteria allow for a great deal of variation to be presented during innovation, in the sense that many employees' and users' opinions are scrutinized (criteria 1–3). Furthermore, the selection mechanisms at the level of the single institution are professional and consumer-oriented (criteria 4–5), rather than just economically. (Purely economic criteria would fall outside our definition of innovation with care.) This is not to say, however, that economic criteria are not important to innovation with care. They can very well be inherent to the two other criteria, especially criterion 5.

We compare the two groups of leaders/institutions – innovation with care and others – on characteristics such as gender, age, seniority, education, type of institution, number of employees, geography and size of municipality in a multiple logistic regression, but there are no significant differences.

In a multiple logistic regression regarding the effects of innovation with care, we find a positive correlation between innovation with care and fulfilment of external demands (see Table 7.3).

In Table 7.3 we see that innovation with care leads to better fulfillment of external demands – two-and-a-half times as often as other types of innovation.

In another multiple logistic regression regarding involvement in strategic decisions we find a significant correlation between type of innovation and extent of involvement in strategic decisions. In the institutions that innovate with care, the leaders' professional skills are more likely to be drawn upon by the political-administrative system in the strategic development.

Table 7.4   Innovation and strategic involvement

Public institutions where the leaders' professional skills "to a certain extent" or "to a great extent" are drawn upon by the political-administrative system regarding the strategic development, by type of innovation and size of municipality. Percentages and estimated odds ratios.

	Involvement in strategic development	
	Percentage	Odds Ratio
Total	60	
Type of innovation		**
Innovation with care	70	2.0
Others	55	1.0
Size of municipality		*
100 000+	50	0.6
0–100 000	64	1.0
N	438	

Source: Danish public institutions. November 2006.
Significance levels: *: $(0.01 < p \leq 0.05)$, **: $(0.001 < p \leq 0.01)$, ***: $(p \leq 0.001)$.

Furthermore, leaders of institutions in municipalities with more than 100 000 inhabitants are less likely to be involved in strategic development than leaders in smaller municipalities (see Table 7.4).

To summarize, institutions which adopt an approach of "innovation with care" have, as reported by their leaders, better chances than other institutions to be in accordance with external requirements and to become part of the strategic development processes in the political-administrative system. In this way, innovation with care at the level of the individual institution (in the domain of innovation) is more likely to be in accordance with and included in the selective processes going on in the broader political-administrative environment (in the field of innovation). In addition to this, this seems to work out in a more successful way in small municipalities rather than large municipalities.

Based on these findings, we now turn to discuss how we can characterize the selection mechanism at the level of the field in the public sector.

DISCUSSION

We have defined innovation with care as a process of innovation that allows for multiple opinions and ideas to be involved in a careful, non-mechanical process of variation and selection. We have also argued that we can distinguish between a domain of innovation, where variation and selection takes place (the individual institution), and a field of innovation where they are selected and diffused across individual institutions.

In the context of the public sector, the domain can be understood as an individual institution such as a school, a kindergarten or home for the elderly. These kinds of institution often belong to a local community and there is little "market-pressure" on them to develop in a similar way based in selection of the best ideas. Although there is often a free choice among institutions, in practice the free choice is restricted. For example, to choose another and "better" service may require more transportation time between home and school.

The field of innovation is the broader political-administrative system in the municipality, the region and the state. In the public sector, selection at the level of the field might be the most difficult point.

On one hand, the field is a place of administrative law and political decision-making as well as the education of professionals, where certain standards and rules for local institutions are set, and where the division of labor among institutions, their size and responsibilities are decided.

On the other hand, there is a tradition of delegating powers to the professionals. Furthermore, public services are in many cases face-to-face services that are dependent on "the moment of truth" in a co-production of the service between the consumer and the provider. The service relation is therefore partly regulated by the personal relationship between the provider and the citizen, although commitment to public rule and a universal service principle is also important.

The question becomes: How do the domain and the field connect to one another? Our survey indicates that public institutions are not deliberately copying one another according to common standards and ideas. It also indicates that institutions that adopt an approach we call "innovation with care" are better connected to the field than other institutions. This means that institutions that engage multiple opinions in their local process of variation and selection seem better to fulfill the criteria in the field of innovation and are better incorporated into the strategy-making of the political-administrative system.

To produce and deliver high quality services based on professional standards and norms seems to be the mantra both for managers and employees. Managers and employees first and foremost have a focus on professional criteria when producing and delivering services and this also seems to be an accepted approach in the broader field of innovation. The result is that a huge variation exists in this field. One could also say that the field of innovation reinforces the creativity at the level of domains, because no strict selection of innovation exists. This again increases the variation of innovation in those fields.

The above does not mean, however, that there is no recognition, reward and selection in the field of innovation at all. It just means that recognition, reward and selection, as in the domains of innovation, are often based on the profession's own standards, norms and traditions for recognition, reward and selection.

To gain a better understanding of what goes on in the selection process in the field, we can make a distinction between two selection mechanisms: convergent and divergent selection. Convergent selection means that institutions (are encouraged to) follow the same standards and that they will eventually follow the same models and do the same things. Divergent selection means that they try to learn from each other and get inspiration from each other.

Our tentative conclusion is that the survey points in the direction of divergent selection as being the most important selection mechanism in the public selector. This can explain why innovation with care is better connected with the field than other forms of innovation. The reason is that there is an acceptance in both the field and in the domain that institutions should try to learn from each other rather than they should implement exactly the same standards based on fixed criteria.

Divergent selection is not necessarily an anarchistic approach to selection. As it has been shown by other case studies in this volume (see Hansen and Serin, Scupola and Steinfield, Møller, Fuglsang), there are many ways in which this divergent selection can be organized.

One example of divergent selection is "the social and strategic arena." For example, a national authority inside a sector, such as the library sector, can undertake a leading role in the funding of projects and selection of ideas that can be used across single institutions. This can also lead to the formation of innovation partnerships among individual institutions, as has been the case in the library sector (see Fuglsang, this volume). Møller (this volume) has shown how the strategic arena comes to play a crucial role for change and innovation in many places in the public sector.

Another example of divergent selection is "institutions for professionals." Research and development in universities and professional schools can give important input to professional development and professional norms. That is, the institutions serve as catalysts for mutual learning and selection that can be explored across single institutions, such as public schools, high schools, social care and so on.

In the above examples, Internet-based information and communication technologies, in the form of portals, social network technologies, or blogs, can sometimes play a crucial role for the expression of new ideas and the diffusion of them across cooperating institutions.

A different example of divergent selection, which may perhaps be more "anarchistic" and less systematic, is media attention, which is sometimes devoted to public services. For example, recent examples of hidden cameras used by journalists who were employed under cover in Danish residential homes in Denmark have put a huge pressure on service development. This leads to criticism of services in both the domain and the field, and to a quest for better variation and selection mechanisms.

However, mechanisms of convergent selection, such as market mechanisms, also exist. An example is consumer choice, which plays a role, for example, among high schools in the Copenhagen area or among universities in Denmark. Consumer choice forces these institutions to adopt some of the same models of education due to competition in order to survive. In some cases, like the library, competition also comes from the outside, in this case the Internet and Google, which leads to convergence between certain information services on the Internet and certain library services. Furthermore, the government creates public rules and standards that have to be followed in a similar way across institutions. Often these standards, however, are seen as minimum standards that allow for variation.

CONCLUSION

This chapter has presented a discussion of innovation with care in the public sector. Innovation with care was defined as an approach that takes care of some of the tensions between variation and selection in innovation, and which allows for involvement and selection among multiple opinions and ideas in a careful and piecemeal way.

We made a distinction between the domain of innovation that is an institution such as a school or home for the elderly where innovation takes place, including local mechanisms of variation and selection, and the broader field of innovation where ideas are recognized, selected and diffused across institutions. This distinction between domain and field was inspired by Csikszentmihalyi (1996).

We defined a proxy for innovation with care at the institutional level, and showed how innovation with care was an approach that was widely accepted. Furthermore, a multiple logistic regression showed that institutions, which adopted an innovation with care approach, were better connected to the field of innovation than institutions that adopted other approaches to innovation. The survey also showed that public institutions in Denmark are innovative and that they do not deliberately copy one another.

We then made a distinction between two mechanisms of selection at the level of the field, convergent and divergent selection, and we briefly discussed different expressions of these. Convergent selection means that institutions tend to copy each other, based on selection criteria that can be grounded in certain common standards and ideas, while divergent selection means that they learn from each other and from the good examples. Our analyses indicate that divergent selection is more common in the Danish Public sector than convergent selection.

Finally, we tentatively discussed different expressions of divergent selection, such as "social and strategic arenas" and "institutions for professionals."

For further research it may be relevant to explore the differences between convergent and divergent selection in the public sector in more detail.

REFERENCES

Amabile, T.M., R. Conti, H. Coon, J. Lazenby and M. Herron (1996), "Assessing the work environment for creativity," *Academy of Management Journal*, **39** (5), 1154–84.

Csikszentmihalyi, M. (1996), *Creativity: Flow and the Psychology of Discovery and Invention*, New York: HarperCollins.

De Vaus, D.A. (2002), *Surveys in Social Research*, London: Routledge.

DiMaggio, P.J. and W.W. Powell (1983), "The Iron Cage revisited: institutional isomorphism and collective rationality in organizational fields," *American Sociological Review*, **48** (2), 147–60.

March, J.G. (1991), "Exploration and exploitation in organizational learning," *Organization Science*, **2** (1), 71–87.

Meyer, J.W. and B. Rowan (1977), "Institutionalized organizations: formal structure as myth and ceremony," *American Journal of Sociology*, **83** (2), 340–63.

Nelson, R.R. and S.G. Winter (1977), "In search of useful theory of innovation," *Research Policy*, **6** (1), 36–76.

Pedersen, J.S. (2007), *Institutionsledere og institutionslederes succeskriterier, rammer, styring og dialog, relation til medarbejdere, tidsanvendelse, innovation & strukturreformen*, Roskilde: Roskilde University Centre.

Popper, K.R. (1962), *The Open Society and its Enemies*, London: Routledge & Kegan Paul.

8. Meta-innovations on strategic arenas: innovative management in public organizations

Jørn Kjølseth Møller

INTRODUCTION

The public sector – including the public institutions – has traditionally been looked upon as being less innovative than private enterprises. This image includes a view of public activities as being rigid and dominated by a lack of incentive for change. Public organizations are viewed as institutionalized and professionalized bureaucracies, where creativity and innovation are limited, and managers and employees are simply expected to carry out the tasks that the political system has defined.

This view of the public sector also exists in the social sciences. Public organizations are typically looked upon as locked into an iron cage, where isomorphism is the norm. According to this view, public organizations are characterized by an increasing rationalization by the use of standardized scripts imported mainly from the private sector.

Even when the focus is on the innovative aspects of public activities, the main notion is that innovation in a public context follows a fixed pattern of change characterized by path-dependence and very little room for maneuver for managers and employees. This kind of exaggerated determinism concerning the potential for development in public organizations is expressed, for example, in neo-institutional analysis and the lack of importance attached to agency as well as in recent theories of path-dependence in welfare policies.

In that way, there has to be a discussion of the possibility of diversity in path-dependence – the occurrence of multiple paths. There also has to be a discussion of the role of institutional agents. This could be the role public employers and trade unions play in negotiations about the institutional context for change and innovation in public organizations (schools, kindergartens and homes for the elderly).

In the following, therefore, the focus is on meta-innovations as a phenomenon – the creation and change of the institutional context to which innovative agents relate, for instance local managers in kindergartens, schools and home for the elderly. Focus is also on some of the characteristics of innovative management in public organizations.

The institutional context and its development can be understood as a kind of negotiated reality which is expressed in a number of "strategic arenas." These arenas define the public organizations as a battlefield of different internal and external interests. These strategic arenas can prompt public organizations to be managed in an innovative way.

One current example of a change of the institutional context in Denmark on a meta-level is the creation of a joint research and development unit for the development of management in local government – "Væksthus for Ledelse," or, in English, "The Management Greenhouse."[1] The two sides of the labor market in the Danish municipalities have created this unit. It is a partnership project dealing with the development of management in which the conception and understanding of what management in a public context is about is negotiated. Focus is also on the scope for change and innovative leadership in public organizations. This is continuously negotiated through activities in concrete development projects which are a crucial part of the "The Management Greenhouse."

INNOVATION IN A PUBLIC CONTEXT

Innovation has become a crucial topic in the debate of what the future will bring societies, such as the Danish, through the challenges of globalization and technological change. Innovation is traditionally understood as the renewal of the condition of production and the exploitation of the potential of knowledge. The aim is to secure and strengthen the position of a society, such as the Danish, in the international division of labor and the economy.

Less attention is given to the consequences of globalization for innovation in the public sector, and its role as a part of the renewal of the Danish welfare society. This is an inconsistency bearing in mind that the public sector represents a vital and important part of the national economy as a whole. With increasing demands on and expectations of public services coming from the citizens combined with politically determined limits to growth in the public expenditure, the need for creativeness and innovation in public activities seems as urgent as in the rest of the economy.

The importance of these issues is underlined by a current reform of the structure of the Danish municipalities. Critical to this reform is the merging of municipalities decreasing their number and, with it, the transfer of tasks and activities from a regional to a local level. The reform is aimed at creating increased economic efficiency, better quality in services, more services and increased democratic governance and influence to the user of the public welfare services (Pedersen, 2004).

A special feature of the production and renewal of services in the public sector is the societal role assigned to the public organizations such as schools and kindergartens. Public organizations are responsible to the public as a whole, they must provide openness in administration, law and order, and public servants are to be impartial, objective and loyal.

This means that the bottom line in every public organization is more complex compared to a private enterprise and, therefore, the condition for the development of innovations are different and often characterized by conflicting criteria of success and failure.

First, public organizations are governed by many political considerations, which means that the political leadership is characterized by competing attitudes and conflicts. This is combined with the fact that the vision for the public sector's development can change with a change in the political majority. Second, the public organizations and administrations are typically organized as bureaucracies dominated by strict rules. Finally, public organizations often carry out a complex mix of tasks ranging from the exercise of legal authority to solving multifarious operating tasks.

Efficiency and quality in solving public tasks therefore have to be carefully balanced and undertaken with careful consideration of the community, democratic values and law and order. With the increased focus on efficiency, effectiveness, competition on a market, documentation of quality and individual adjustment of services combined with the classical virtues of law and order, stability and continuity, public organizations and their employees are therefore to an increasing extent placed under conflicting pressures that complicate creative and innovative thinking and action. At the same time this also underlines the requirement of innovative activities.

NEO-INSTITUTIONAL ANALYSIS AND INSTITUTIONAL AGENTS

Since the early 1990s, a number of analyses have shown the path-dependence of public policies and socio-economic institutions (Hodgson, 1994). These analyses emphasize the dependency of social agents on specific institutional set-ups limiting the scope for maneuvering and change in vital economic, political and societal areas. By the very choice of a specific institutional set-up, the potentials for institutional innovation and change seem already to be limited, or nearly nonexistent, according to this approach.

According to the neo-institutional tradition in political science and sociology, all agents therefore seem to be placed in an "iron cage" from which they cannot break free (Powell and DiMaggio, 1991). This view seems, however, to be contrary to the everyday experiences of pro-active institutional changes that are reported in many parts of society. We are therefore missing an understanding of how public institutions change. Changes do take place in regulations, norms and cultural-cognitive aspects of public institutions (Scott, 2001), but they are badly conceptualized by institutional theory as defensive changes leading to homogenization. One example of how institutional theory views institutional pressure in, for example, education is shown in Table 8.1. A proper theoretical perspective which gives the agents the potential for action – creating innovations and organizational change – is therefore simply missing in a neo-institutional perspective on public organizational cultures.

By introducing a concept of institutional agency, a potential for change is created. The agent is characterized by the urge to seek new solutions, which have not earlier been viewed as relevant and possible in the existent institutional structure. These innovative social agents open up the possibility for change by combining known elements of the institutional structure in a

Table 8.1 Institutional pressures on educational organizations

	Regulative pillar	Normative pillar	Cultural-cognitive pillar
Institutional structure	Explicit rules	Values and norms that are informal and internalized	Picture of the world and collective idea
Institutional impact	Formally accepted or forbidden types of behavior	Given standards for right and wrong behavior	Creating environmental and own pictures embedded in persons/groups, in such a way that specific behavior is regarded as sensible
Sanctions when breach of rules	Punishment	Exclusion from the community (impose guilt)	Being characterized as unwise and irresponsible
Institutions, examples	Law about kindergartens	Non-written rules of the educational profession	Analyses by the social partners defining management skills

Source: Kjær (2006).

new way, when possible. The notion of the institutional agent refers to a broad range of political and economic agents at national and local levels. The institutional agents could, for example, be employers' organizations, trade unions, organized users, political parties or major public and private organizations and enterprises, which each dominate their specific field of organization or domain.

The way the institutional agent influences the development of innovations is by *meta-innovations*, that is by changing the institutional context in which the innovator (an innovating organization or R&D person) is typically operating (Crouch, 2005).

One example of institutional agency and meta-innovation in Danish municipalities is the employers and trade unions who, in 2004, created the The Management Greenhouse. The aim was to support the development of management in the new regional and municipal structure after the reform of municipalities and regions. By combining the traditional concern for negotiating wages and improving working conditions with a new concern for development and change, the social partners changed the institutional context for Danish municipalities. The understanding of a mutual dependence – a kind of resource dependence (Pheffer and Salanik, 2003) – between the social partners has driven this change, together with a joint uncertainty about the prospective consequences of the current municipal reform (which represents the largest overall organizational change in Danish society for several decades).

INSTITUTIONAL DEVELOPMENT OF THE DANISH SYSTEM OF MUNICIPALITIES

The development of the Danish municipalities system has for the last four decades been dominated by four emerging reform strategies with one overriding theme: decentralization of the public organizations and, thereby, the creation of a changed institutional structure (Klausen, 2001).

The first wave of renewal and reform from the end of the 1960s through the 1970s incorporated the transfer of tasks from the state to the local level of municipalities together with a reform of the structure of municipalities in 1970. This first wave of changes involved an expansion of the welfare state system, a professional production of welfare services, flat structuring of organizations and joint and collegial management, and management by rules and loyalty connected to each working place.

The second wave of reforms in the 1980s involved a decentralization of economic responsibility to the local management of the individual organizations, a slowing down of the increase in local welfare state expenditures, a focus on efficiency and the development of quality in services, empowering the local managers, the implementation of new principles for local governance and an increasing awareness of a joint destiny in local government.

The third wave of reforms, which took place in the 1990s, saw the introduction of experiments with the outsourcing of the municipal activities (New Public Management), Total Quality Management, centralization and even more hierarchy, the use of contracts (between municipality and the single public organization) and management across the different sectors of local government.

The fourth wave of reforms, after the millennium, has been dominated by the present reform of the whole system of local and regional government with the aim of rationalizing the municipal structure transferring tasks from the regional to the local level and the merging of individual municipalities. Economic efficiency, political effectiveness, the development of quality of services and the empowerment of the users have also, together with a focus on values, social networks and loose coupling, management as a profession, centralized strategic governance and decentralized responsibility of the management and the integration of the individual municipal organizations in the broader municipal community dominated the agenda. The aim of this integration has been to increase the sense of being together locally, and to close gaps due to fragmentation of public organizations and the welfare state.

In this way, the whole development of the system of Danish municipalities during the last 40 years represents a kind of institutionalization "layer on layer" (the four waves of reforms) having the following four dimensions: tasks, structure, culture and governance modes. Even though the common denominator or theme of this institutional development has been "decentralization," the imprint has been set on many different paths and development traits. It is remarkable that in relation to the planning of the present reform of the system of municipalities more than 27 different criteria of successes have been on the political agenda (Pedersen, 2004). The most important criteria have been: increased economic efficiency, more quality, more services and increased democratic governance and user influence.

In this light, the need for innovation in the activities of the municipalities must be connected to the ongoing overall renewal of services (tasks), structure (management and work organization) values and understanding (culture) and governance modes (the re-combination of hierarchy, network and the market as a governance mode).

One striking feature of the development of the Danish – and in this respect also the whole Nordic – welfare model has been the "innovative translation" or "creative" inventiveness as an answer to the international reform pressure and trends. An illustration of this is that the wave of privatization and outsourcing during the 1980s was translated to the "modernization" of the public sector, and New Public Management during the 1990s was translated to the "delegation of economic responsibility" to the local level in the municipal system of institutions.

The main point is, therefore, that in reality there is a large amount of diversity in path-dependence across both national borders and in different local cultures in, for example, the system of Danish municipalities. This diversity has recently been documented in a report from a research project financed by The Management Greenhouse on the development of management structures for educational organizations (kindergartens) in Danish municipalities (Møller *et al.*, 2007). It is this diversity that, to a certain extent, creates the possibility of renewal and innovation in public organizations (too much diversity may, however, destroy this potential for innovation by reducing the learning effect across individual organizations).

The wave of privatization has therefore been very different in for example the UK and the Nordic countries, and the extent of privatization of municipal organizations and services is likewise very different across municipalities in Denmark. There are not few but many ways of developing public organizations depending on the cumulative effect, the implementation of new management principles and governance modes. A path or trait is established to the extent that the imprint is being set on the path, but can as a potential path also disappear (become overgrown), and even afterwards be found again, and the imprint then set on it again.

STRATEGIC ARENAS IN PUBLIC ORGANIZATIONS

The strategic issue – the need for innovation – in the public sector in general, and in the municipal activities and organizations more specifically (for example schools, kindergartens and homes for the elderly), can be looked upon through the concept of "strategic arenas."

A strategic arena represents an environmental dimension or strategic field in private and public organizations, where specific types of interests are formulated, specific issues negotiated and specific rules of the game are defined. They reflect vital parts of the business environment that the management has to consider in its attempt to carry out renewal and innovation (Møller, 2005).

In brief, one way to categorize the different strategic arenas that may exist is in a typology that contains eight strategic arenas with different focal points (Klausen, 2004):

1.   The arena of production. This includes the organizational conditions for the production of services and is linked to the way it is organized and managed.

Table 8.2   Examples of strategic arenas in educational organizations (kindergartens)

Strategic arena	Definition
The arena of production	The "technical" condition for the delivering of educational services such as the organization of the educational work, pedagogical knowledge and working methods
The social arena	The social environment around the organization of the educational work and the socialization of the employees
The market or the domain for educational services	Who is supplying educational services (care for children) and who is demanding educational services, and what are the competitive conditions (for example public or private cares for children)?
The arena for political decision	The power struggles and the rules of the game, which characterize the environment of the educational organization, where decisions are made in a competition, an interplay and a negotiation between different interests (for example institutional agents)
The arena for knowledge and sense-making	The "discourse" in the society of what an educational service is (for example daily care for children, socialization, education and learning), and what is the strategic situation and problematic of the domain
The arena for visions and basic assumption	The conception of what public educational services is part of (for example the importance in relation to the development of the welfare state)
The cultural arena	The cultural environment shapes the values which govern the behavior of the employees, and the values which ought to govern the behavior
The arena for architecture and aesthetic	The physically designed environment, which create the frame around the daily educational work in the educational organizations. For example the physical and aesthetic design of the playgrounds and playhouses

Source: Klausen (2004).

2.   The social arena. This deals with the working environment of the employees and the social life of the organization.
3.   The market or the domain of the organization. This concerns the market or domain in which the organization operates and competes. Who are the competitors?

4. The arena for political decisions. This concerns which kind of decision-making bodies and processes the organization depends on. Who are the interest groups or agents and what is their power base?
5. The arena for knowledge and sense-making. What are the dominating knowledge base and understanding in the organization? How is the "world inside and outside" structured and looked upon, situations understood and interpreted.
6. The arena for visions and basic assumptions. What is it that the members of the organization believe in and what are the visions for the future of the organization.
7. The cultural arena. Which kind of core values are guiding the behavior in the organization?
8. The arena for architecture and aesthetic. The way the organization appears and functions physically and aesthetically (the artifacts).

Each of these strategic arenas contain their own specific problematic, which typically has a number of theories and models (Mintzberg *et al.*, 1998) which can be used to shed light on different aspects of these problematics and making it possible to manage the organizational development through different innovative strategies to handle the challenges of the problematic.

They also represent different battlefields, where "friends and enemies" struggle on different issues on the basis of the rules of the game, which determine the struggle in this specific arena. What the struggle is about in a public context is legitimacy and support to be able to carry out dreams and needs and thereby getting admission to vital resources (Pheffer and Salancik, 2003). It is, however decisive, how this legitimacy is created and conquered, for the support and gain to be real (Suchmann, 1995). It is in this relation that managers of public organizations show their strategic leadership.

PUBLIC INNOVATIONS AND SYSTEMS OF INNOVATION

Koch and Haukness define innovation in the following way:

> Innovation is a social entity's implementation and performance of a new specific form or repertoire of social action that is implemented deliberately by the entity in the context of the objectives and functionalities of the entity's activities. (Koch and Hauknes, 2005)

This definition of innovation focuses on innovation as a behavioral expression of the public organization's intentions and goals, which are shaped by the understanding of the social context, the technical and institutional environment of the members of the organization.

In this understanding innovations are diffused through the sharing of knowledge and learning inside and across organizations by the way members of organizations understand and accept them. There is in this way a close relationship between innovation and learning.

The definition also implies that innovations must represent something new for the public organization, for example a change in behavior, which is new to the organization in their choice of action repertoire, but not necessarily new to society as a whole.

BOX 8.1 TYPES OF INNOVATIONS AND INNOVATION STRATEGIES IN THE PUBLIC SECTOR

Product and process innovations:

- A new or enhanced service (for example care for children).
- New working processes (for example changes in working routines at a home for elderly).
- Administrative renewal (for example political-administrative governance modes, which create decisive policy change in relation to the local institutions).
- System change (for example reform of the structure in the system of Danish municipalities, which creates larger entities, new patterns of co-operation, and a changed and better organization of the production of public services).
- New concepts (for example a new understanding of the management task, network as a management strategy, the use of internal contracts).
- New rationales (for example that the services of kindergartens is not only about daily care for children, but also is a way of integrating children of emigrants in the Danish society).

Innovation strategies:

- Incremental versus radical innovation (for example gradual extension of existing public services against radical new public services).

- Top down versus bottom up innovations (for example the question whether changes in management structure of educational organization is implemented from above by top management or from below in a cooperation with the local managers of the public institutions).
- Innovations determined by needs versus by focus on effiency (for example whether innovation is directed at solving a specific problem or directed at a more efficient use of resources in the public organizations).
- User versus production driven innovations (for example whether new educational services for children are aimed at solving parents specific needs for care of children or the organizations need for more rational working processes).

Source: Koch and Haukness (2005).

Traditionally, the employees of public organizations have been regarded as subordinated employees in relation to politicians and top management. The employees have not been expected to be able to or permitted to put forward new ideas on how the public services could be changed and renewed, but only to deliver the services which were decided politically.

Although many public employees still find it difficult to be innovative in their working role, because their role in the public system as supplier and producer is highly institutionalized through years of practice and tradition, this pattern is being changed gradually in relation to the decentralization and fragmentation of the public sector.

There is, as such, a growing focus on the need for a new kind of "entrepreneurship" among the public employees and specifically among the local managers in the public system. It is accepted and even increasingly expected that the managers take the initiative and become more creative and innovative in the development of the public institutions. This development is also driven by the increasing complexity of the public tasks and problem solving, which means that earlier hierarchical governance modes are not so usable any longer. As it is argued by the PUBLIN project (Koch and Hauknes, 2005) the role as an "entrepreneur" is not far away from the tradition and virtues of the dominating professions among the public employees and looks almost like the role of some of the political and institutional entrepreneurs, because they are:

- Ideological minded to be innovative. They have a picture of the world that gives them the necessary belief in the need for change.

- Idealistic or altruistic. They have chosen the public sector, because they want to be part of social change.
- Aware that being creative and innovative they can promote their own carrier. For some of the public managers, this also implies appearing as "institutional entrepreneurs" trying to move the limits for what is permitted and accepted in public activities.
- Filled with intellectual curiosity and stimulated by their wish to see the challenges in the need for change. Typically, this includes people who have had a higher education with a great deal of professional self-assurance.

As pointed out above, the increasing complexity and fragmentation of the public service also means that there is a wider space for innovative activities in public organizations. It is nearly a political and administrative demand.

Given the need for creativity and innovation in the public sector, it could therefore be useful to try to map the social factors which stimulate innovation and creativity, and the factors representing barriers to new ways of delivering public services (see below).

In order to deal with complexity, fragmentation, innovation and learning, the agents in the public arena must enter into a close interplay with one another. Their ability to become innovative depends on their ability to find relevant competencies, understand them and use them in their own context. Therefore the creation of innovative networks among managers of public organizations is also a vital aspect and basis for the system of innovations in the public sector.

A network can be formalized in differing degrees (Møller *et al.*, 2006), and serve different goals depending on the intentions. A network can be used to solve "small" problems through interchange of experiences, but also be part of a more systematic innovative process where the participants aim at implementing a more substantial innovative change in the public production of services. The latter is critical to new municipal network of managers in Denmark supported by different programs initiated by The Management Greenhouse. Managers in networks therefore become a management strategy in the municipalities, which support and fill in the need for change across the municipal organization.

Seen as an institutional network for learning and innovation – and thereby as a public system of innovation – the many new networks of managers can be understood by using the model represented in Figure 8.1 (Koch and Hauknes, 2005).

| The ministry of family (agencies) | Local government | Municipal administration | Interest organizations (KL/BUPL) |

| Private service suppliers (cleaning) | | | Other institutions for children (schools, libraries) |

| Private technology suppliers (IT, toys, tools for playing grounds) | ↕ Educational organizations. Supplier of educational services. Coordinator of innovative activities. ↔ | Media and the public |

| Private consultants (management) | | Users, clients and citizens |

| R&D departments of companies | Universities | Institution for education of the professionals (CVU's) | Sector specific research institutions (DPU) |

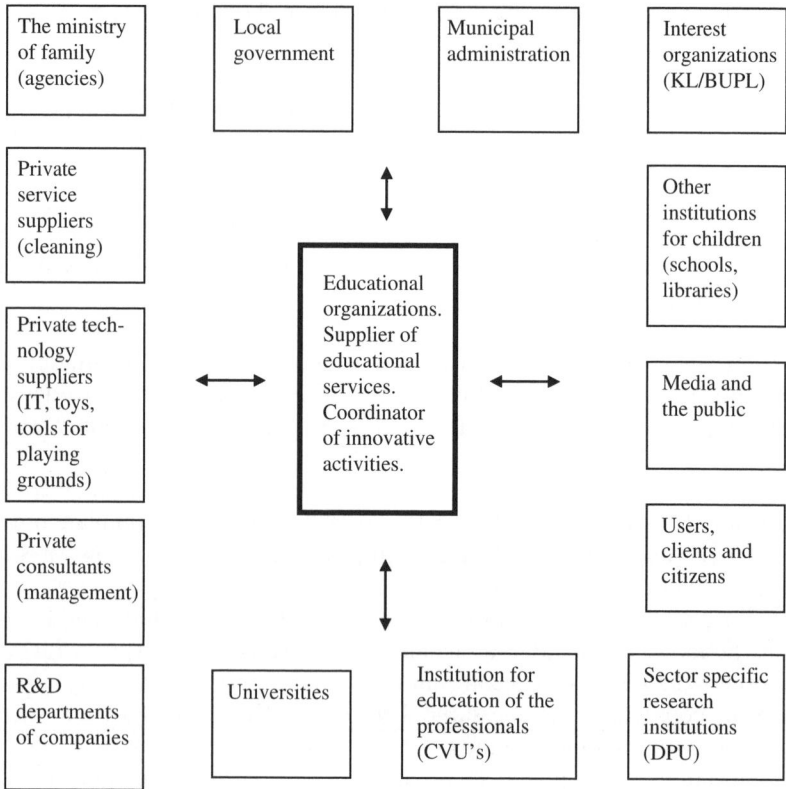

Source: Inspiration from Koch and Haukness (2005).

Figure 8.1 System of innovation on the domain for educational services in Denmark (kindergartens and services for care of children)

DRIVERS AND BARRIERS

Innovation appears in many different connections and forms in the public sector. In general, innovation can be divided into what is going to be innovated, what forces are driving innovation and how innovation happens.

In a public context, compared to, for example, industrial innovations, there is a high degree of renewal in services, tasks and processes rather than in the development of physical products. Innovation in services covers a wide spectrum from quality in services for elderly people to electronic services for the citizens, and the processes of the simplification of working routines, use of working methods and changed methods of management. But the democratic modes of governance are also covered by process innovation as it happens with the structural reform of Danish municipalities. Even the shift in mental models relating to the understanding of public organizations and their primary tasks can be defined as an innovation.

Innovation in public organizations is driven by user needs, ideas of employees and by their competencies, new knowledge, political intensions and the development in prices and expenditures (less expensive ways of producing the same services). Lastly innovation is created by combining known and new knowledge in a new way or put together in a new connection.

Drivers are forces that put innovation in motion – a development or a change. Innovation can be forced by pressure from outside for a change in the organization, but it can also be connected to internal issues and unhappiness with a given way of dealing with tasks.

- Pressure from outside. When public organizations are put under pressure from something happening in the environment, it opens a possibility for new methods to be implemented or developed or for new services to be created. The reason could be a cut in budgets that challenges the public organization to find creative solutions to be able to keep up the size and quality of services, or a shutdown that forces the public organization to radical new ways of thinking. The experience is that when the drivers come from the outside, the innovation is often initiated by the top management.

- Internal problems. When something is experienced as an irritation, a returning problem or challenge that does not find a solution, or one cannot move, then there is a possibility that something new can happen.

- Motivation for improvement. A huge part of the public sector is dominated by strong professional groups with high professional aspirations and ideals. Many public employees are both ambitious in a professional and a social sense, and therefore they are also eager to set the agenda for the development of the welfare state. Professionals can therefore on their own be drivers for renewal or innovation in the public organization.

- Coincidence. Innovation is often not planned, and the results are not known beforehand, therefore coincidences will often be the impetus to innovation. When employees and their managers are surprised or they get a question that they cannot answer, there is a motivation to explore it and new ideas may emerge.

But often there are also a number of *barriers* to innovation in the public sector. In general, innovation is a dangerous project to participate in – both personally and professionally. Innovation deals with ideas translated to a better praxis. To be innovative you have to be involved in a project of change where it could be a question of changing bad habits or the distributions of resources and tasks. The most important barriers for innovation in public organizations are the following:

- The size and complexity of the organization. Size and complexity can be a major barrier for innovation in public organizations, because the public bureaucratic organizations so characteristic for the public sector both have the responsibility of producing services and at same time of being a legal authority – all under democratic control. It gives a fixed agenda for many public organizations.
- Bureaucratic organization for it own sake. Bureaucratic organizations are characterized by a clear distribution of responsibility and tasks. That is why they are efficient according to Max Weber (Weber, 1976, Part III, Chapter 6). But bureaucratic organizations are not the organizational form that really supports employees to risk experiments in problem solving or dealing with tasks outside their given responsibility and competencies. Therefore new forms of organizing the public activities such as networks are used to an increasing extent.
- Demarcations and professional identity. Professional cultures can be a driver of innovation in public organizations, but the professionals can also be seen as a barrier because of their monopoly of knowledge. Their professional identity is often connected to defensive routines, working methods and praxis, and therefore they have a tendency to look upon ideas from outside as a threat to their professional autonomy. This is also the problem when other professionals are moving into their professional domain.
- Lack of systematism. There is not a tradition for systematizing either the user driven or the employee driven innovations. A lack of a systematic development of an innovative culture in public organizations can be a barrier for the exploitation of innovations.
- Laws and rules. There are many rules and bureaucratic routines in the public sector and every time rules or routines have to be followed nearly the opposite of innovation occurs. There is therefore often a great opposition to risk taking or even more to showing a kind of civil disobedience.
- Lack of innovation strategy. Many municipalities and public organizations have no innovation strategy or even an innovation budget. When the general strategy is missing – while a lot of developmental projects are started – there is the risk that a form of innovation fatigue will emerge.

- Short time horizons. A barrier for innovation is the fact that innovations take time, that is much longer than is typically within the planning horizon of many public organizations. It is therefore a crucial part of the creation of an effective innovative culture that the innovative work is based on a long term planning horizon that supports the efforts to develop improvements and innovations.
- Implementation. When the proper amount of resources to secure the transfer from idea to practice is not allocated, it is often the case that nothing happens. The problem is not to generate new ideas, but rather to try to implement them in the organization.
- The expectation of knowing the result beforehand. If the public organizations believe that creativity and innovation can be planned and steered in detail, then the result is often negative and the organization ends up not to be where it planned to be. The innovation then becomes unsuccessful. Innovations require that there is space for experimentation.
- Shortage of competencies. Often managers as well as employees in public organizations express that they need more knowledge about how to create innovations. There is a need for education in creativity and risk management and knowledge too about how in practice to learn to get rid of old habits and routines that can be a barrier for the new to happen.

INNOVATIVE MANAGEMENT IN PUBLIC ORGANIZATIONS

The management of innovation or innovative management is what managers do to support ideas being generated and being tested and followed by action, assessed and systematized in practice. Innovative management is a question of supporting and creating the frames for innovation – independently of who is doing what. The management of innovation involves a holistic view and long term goals and visions. It establishes a developmental view, seeing innovation in the light of the values of the organization. The management of innovation is about prioritizing and choosing among alternatives.

Sometimes it is primarily the top management of public organizations that carries out innovative management, but in reality it is more often the employees in co-operation with the local manager that manage innovation. The reason is either that the employees prefer to manage themselves or because this is the way it is.

There are different ways of managing the innovation efforts. Some are long term and indirect. The task could be to interpret and influence the external condition for innovation or to decide the frames in the day-to-day work with innovations. This is a structural approach to the innovation task. In this relationship, the management is focused on the strategic level and developing the innovation culture.

Other forms of management of innovation are shorter term and thereby also more direct. Management is undertaken close to the innovative activities, and the process is being organized by the manager and employees are receiving coaching on a regular basis.

The management of innovation therefore implies that the manager is involved in one or more of the roles in Box 8.2.

BOX 8.2   FIVE ROLES IN THE MANAGEMENT OF INNOVATION

The strategic agent
The manager focuses on the general conditions and works on the long term aim to create the optimal frames for innovation.

The developer of culture
The manager works on the long term aims with the internal innovation culture by acting as a role model and by stimulating specific activities supporting the culture.

The team builder
The manager recruits innovative employees and combines employees which in an interplay with others contribute to increase innovation.

The distributor of tasks
The manager deals with the coordination of tasks and activities whereby innovation becomes a natural ingredient, securing for example a proper amount of challenging tasks combined with activities in networks where the new solutions are spread and experiences interchanged, and given the possibility of improving chosen solutions.

The coach
The manager is personally involved and close to the activities. He is the challenger, asking questions and the creator of ideas which improve the efforts to innovate.

Source: Digmann (2006).

DIVERSITY AND THE CAPABILITY TO INNOVATE

There is a relationship between diversity in organizational culture and innovative capability, that is the ability to generate new ideas and implement them – both in the individual organization and in municipal organization as a whole. Diversity is a limiting factor if there is too big a conflict between different subcultures in the organization in relationship to what is the main task for the organization. This is because it limits the possibility of communication and knowledge sharing in relation to the use and combination of known knowledge in new connections. Similarly, a dominant culture, which creates conformity and group thinking among the employees of the organization, will limit the possibility of the organization of choosing other directions than those which follow from the dominant culture.

Similarly, the organization in a municipality will, if it is fragmented and divided into different subcultures, limit the possibility of a transfer of knowledge from one part of the organization to another. The dominant culture and mode of governance can cut off the possibility of seeking other solutions than the ones that follow from the official and authorized type of management.

The relationship between the dominant culture, diversity and capability of innovation is illustrated in Figure 8.2, where the dominant culture is characterized by low or no diversity.

From a neo-institutional perspective it could be said that there is a similar relationship between path-dependence and the ability to create renewal and innovation in the chosen institutional set-up.

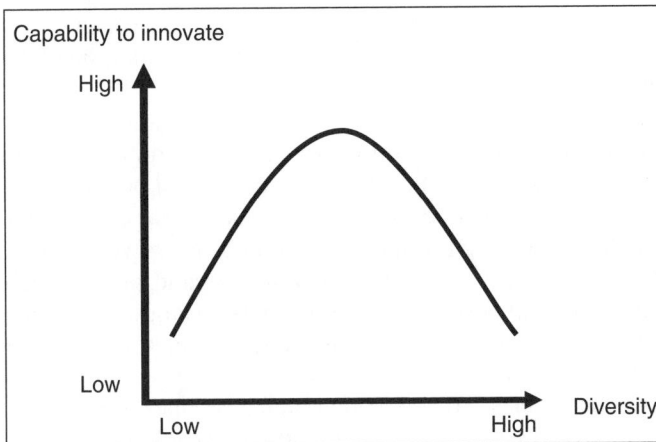

Source: Inspiration from Klausen (2004).

Figure 8.2 Diversity and capability to innovate

INNOVATION AND LEARNING IN A NETWORK

Networking has been put on the agenda as a mode of governance in Danish municipalities, because the understanding of what public governance is and ought to be has changed remarkably in recent years.

The background is the increasing complexity and the dynamics of the public sector with an increasing fragmentation of the public organization and therefore an emerging co-ordination and steering problem.

The known mode of governance is no longer sufficient. It is built on the understanding of a coherent parliamentarian steering chain which goes from local government to the management of the individual public organization with the municipality's top level management placed in between. However, it has to be complemented with structures and mechanisms, which secure coordination and self-management on a local level in the municipal organization. A form of self-management which has to be consistent with general policies and strategies of the local government and important issues on the political agenda.

The use of networks of managers as a part of the management structure has the potential for such a type of self-management and represents one among several answers to solving the problem of integration and coordination in the municipal organization.

But a network of managers as a co-ordinating management mechanism cannot stand alone. It depends on a clear and explicit strategy from the political leadership and the top management which describe the direction and frames for the development of the public services. In relation to this, the top management must formulate some core values which must characterize the public service organization, and a set of management principles for the local manager.

In a modern western society, the dominating organizational culture is now changing from a role culture with rules, procedures and formal job descriptions, and a power culture built on power and position, to cultures more focused on person and tasks. This is characterized by a high degree of autonomy concerning the choice of methods for solving tasks, and the relation to the customer or citizen. Knowledge and professionalism are valued higher than power and position. Charles Handy has described this change in terms of four different organizational cultures, see Figure 8.3 (Handy, 1985).

The creation of networks of managers across the organization enables this change to organizational cultures focused on person and tasks. It gives, as mentioned above, the possibility of both a planned and co-ordinated development of the public services, the creation of a consistent management practice – a common pattern – on the local level across the different parts of

Power culture   Task culture

Role culture   Person culture

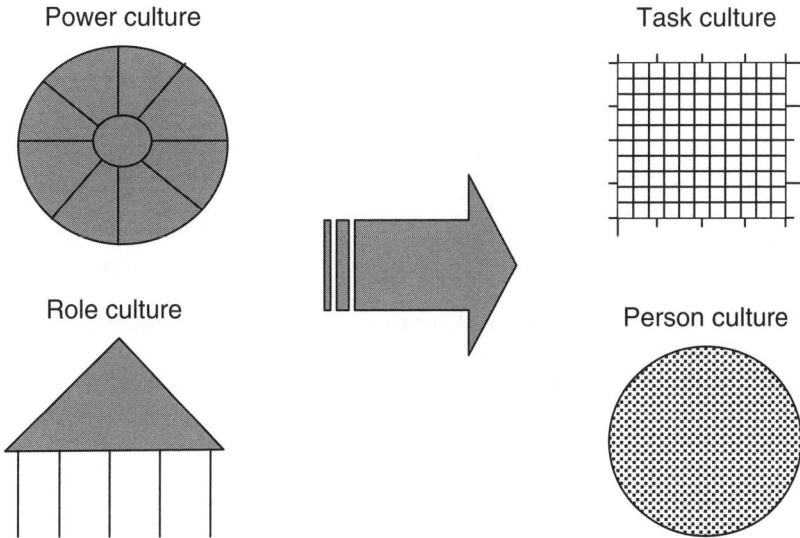

Figure 8.3   Organizational cultures

the public organization. Furthermore, it enables knowledge sharing through a common learning in the networks, and a positioning and selection of those services, which by their example can counteract an inappropriate mutual competition between the public organizations, or competition between areas of public services, for scarce public resources.

In addition, the network can facilitate a shared view on how to develop the public service organization and a constructive interplay between the top management and local management based on dialogue about the future development and implementation of strategies. A well-run top management is characterized by a leadership, which sets the direction, focuses the efforts, defines the organization and creates a consistent pattern of solving tasks in the public service organization.

If the networks of managers have to change from a state of having a potential for co-ordination and learning structure to a state of active co-ordination and learning, experience shows that the network must change from being unfocused and created nearly by accident to becoming a focused and task oriented network.

A way of promoting the capability of innovation even in a fragmented public organization with many subcultures is by knowledge sharing in a social network. Especially important could be a network that is purposeful in promoting an innovative culture across the organization. It is typically

Table 8.3   Characteristic of networks as a potential and as action (intention)

	Network as a potential	Network as action (intention)
Drivers	The participants	The task
Being physically close	Important	Not important
Crucial part	The individual	The task
Management	Coincidently	Design of the task/ recruitment
Result for the individual	Unstructured knowledge and learning	Structured knowledge and learning
Result for the organization	Unclear and accidential	Focused and specific (solutions)
Development of competencies	A potential (need action)	More competent action
Development of network	Coincidently and non-systematic	Increase in complexity of tasks, communication and change in organization
Strength	Exist everywhere and is flexible	Demand for results create the drive and meaning (increase in values)
Weakness	Non-focused and no responsibility	Takes time, engagement and responsible participation
Conditions	Personal interest, time, acceptance	A task, infrastructure, priority and awareness
Range	Unlimited	Defined by the task and the potential of the network
Organizational cost	Depends on whether the participation is part of daily working time	Use of working time, but the development of competencies happens through the work (the task)

Source: The scheme is an adjusted scheme based on Connie Svabo, in Larsen and Svabo (2002).

on the border between different communities of practice that learning happens and innovation is created (Wenger, 1998). The present period of reform in the system of Danish municipalities networks has, therefore, become a vital mechanism for sharing new ideas in many Danish municipalities (Møller *et al.*, 2006).

Municipal networks of managers represent a forceful way of creating learning and develop competencies – both on organizational and individual levels. "Competencies" means, in this connection, the ability to carry out competent action in the public organization.

Networks represent, in this relationship, both a space for potentials and actions, but they only work when used through shared activities and actions. The word "networks" contain the word "work." Net*work*s therefore only work when there is activity.

As mentioned above, it is the focused and task oriented networks which are the most efficient to create competent action in public organizations.

THE MANAGEMENT GREENHOUSE AS A META-INNOVATION

In 2004, the two parties on the municipal labor market – Local Government Denmark (LGDK) and Komnunale Tjenestemænds Organisation (KTO)[2] – created a joint organization for development of management to promote the development of new and innovative leadership in the municipal organization.

The official goals for The Management Greenhouse are:

- a systematic and purposive dialogue on management and with a focus on management skills;
- a forum for debate about the future municipal workplaces and working processes;
- a development track that, together with the negotiation track, can develop co-operation and working processes in the municipalities and regions;
- a promotion of experiments and projects of development that can try out new forms of management and co-operation;
- a sharing of knowledge and promotion of the work done in the greenhouse.

In the 2005 collective bargaining round, the parties decided to allocate a total amount of DDK 25 million to be spent over the next three years to fulfill the goals of The Management Greenhouse.

The organization of the Greenhouse is divided into three smaller greenhouses, namely one for top management, one for middle management and one for local management at the level of institutions. Each greenhouse has it own resources for development.

During its first years the Greenhouse has initiated between 15 and 20 different projects, which cover for example a strategic workplace for top management, the successful local manager, a learning project for innovative

managers, the education of consultants helping managers creating networks, Lean Management as a management tool and new ways of organizing the management of educational institutions. All the projects are
implemented as co-operative projects with participation from both sides
(LGDK/KTO) in the greenhouse. There has also been a debate about public
governance (a codex for good management in Danish municipalities). The
projects and their results are documented on www.vaeksthusforledelse.dk.

The Management Greenhouse can be looked upon as an example of how
two institutional agents and co-operating parties on the labor market
(LGDK and KTO) create a form of a "meta-innovation." They combine
the traditional track of wage negotiation with a new track of development
of management. By building a new institutional framework or context,
where management and renewal of the public organization can be tried out,
the parties influence the way the management practice will develop during
the coming years. Most of the projects initiated by the Greenhouse incorporate one or several of the strategic arenas for local management mentioned above.

The reason for the extended co-operation, and thereby the contribution
to the institutional development in the system of Danish municipalities, is
without any doubt a shared recognition of the inter-dependence between
the employers and employees in the present state of a major municipal
reform. If the reform is to be successful and develop in a socially balanced
way and in accordance with the officially declared intensions, there needs
to be a good relationship between the two parties over the coming years
in which the organization of the municipalities will be changed. At the
same time, the new way of co-operating is also an expression of a joint
uncertainty about the future consequences of the largest organizational
change project in Denmark for a decade or more. The joint efforts in the
Greenhouse is also a way of securing peace in both "political" camps in
relation to this very complex project in the whole system of regions and
municipalities. Anyway, the Greenhouse represents a major and innovative
institutional change in the Danish labor market with consequences for
management practice and the relationship between management and
employees in a vast part of the Danish labor market.

CONCLUSIONS

The possibility for innovative management in public organizations is
determined by the institutional context, where the public institutions are
functioning. A number of strategic arenas are critical for this. The strategic
arena is an institutionalized environment to the organization, where specific

types of interests are expressed, specific issues negotiated and specific rules of the game established.

The development of the institutional context is not determined by a single, but many different development tracks, which, as time goes by, become embedded "layer on layer" in the institutional structure relatively independent of the present mode of governance, existing values and norms and the prevailing cultural-cognitive understanding.

The Management Greenhouse is an example of a meta-innovation. By creating a new institutional context for renewal and innovation in management practice, incorporating several strategic arenas, the parties on the labor market act as institutional agents or entrepreneurs; they create a meta-innovation (the Management Greenhouse) which will influence the individual organizations in the coming years.

The concept of institutional agents can be used to represent a much broader range of agents than those mentioned above. It refers to both organized interest groups, such as trade unions, and also to private and public enterprises and institutions that appear as institutional "entrepreneurs" when they, in their daily practice, try out new possibilities and move the limits for legitimate behavior. Institutional agents can change the institutional landscape, thereby creating a new framework for what is socially and politically accepted in society.

NOTES

1. The author of this article has participated as project manager in The Management Greenhouse since 2004, when the organization was created. He is presently also an industrial PhD student.
2. Local Government Denmark (LGDK) is the association of the Danish municipalities and Kommunale Tjenestemænds Organisation (KTO) is the association of trade unions in the Danish municipalities.

REFERENCES

Crouch, C. (2005), *Capitalist Diversity and Change – Recombinant Governance and Institutional Entrepreneurs*, Oxford: Oxford University Press.
Digmann A., H.W. Bendix, J.P. Jensen and K. Jensen (2006), *Offentlig innovation – I balancen mellem idé og systematik*, Copenhagen: Børsens Forlag.
Handy, Ch. B. (1985), *Understanding Organizations*, Harmondsworth: Penguin.
Hodgson, G.M. (1994), "Lock-in and chreodic development," in G.M. Hodgson, W.J. Samuels and M.R. Tool (eds), *The Elgar Companion to Institutional and Evolutionary Economics L-Z*, Aldershot: Edward Elgar.
Kjær, P. (2006), "Institutionel teori til analyse af strategizing," in C. Nygaard (ed.), *Strategizing – kontekstuel virksomhedsteori*, Copenhagen: Samfundslitteratur.

Klausen, K.K. (2001), *Skulle det være noget særligt – Organisation og ledelse i det offentlige*, Copenhagen: Børsens Forlag.
Klausen, K.K. (2004), *Strategisk ledelse – De mange arenaer*, Odense: Syddansk Universitetsforlag.
Koch, P. and J. Hauknes (2005), *On Innovation in the Public Sector*, Oslo: Publin Report no. D20, NIFU STEP.
Larsen, H.H. and C. Svabo (2002), *Fra kursus til kompetenceudvikling på jobbet*, Copenhagen: Jurist- og Økonomforbundets Forlag.
March, J.G. and J.P. Olsen (1989), *Rediscovering Institutions – The Organizational Basis of Politics*, New York: The Free Press.
Mintzberg, H., B. Ahlstrand and J. Lampel (1998), *Strategy Safari – The Complete Guide Through the Wilds of Strategic Management*, Harlow: Prentice Hall.
Møller, J.K. (2005), *Strategisk ledelse og forandring*, Vejle: Kroghs Forlag.
Møller, J.K., Flemming Blønd, Torsten Petersen, Jesper Mathiesen, Sanne Brønserud Larsen and Mette Seier Helms (2006), *Ledere i netværk – En ledelsesstrategi*, Copenhagen: Væksthus for Ledelse.
Møller, J.K., Camilla Hagelund, Dorthe Nørgaard, Lise Balslev and Vibeke Schneider (2007), *Styrket ledelse gennem nye ledelsesstrukturer*, Rapport fra undersøgelsesprojekt om Ledelsesstruktur og lederfaglighed på dagtilbudsområdet, Copenhagen: Væksthus for Ledelse.
Nielsen, K. (2005), *Institutionel teori – En tværfaglig introduktion*, Frederiksberg: Roskilde.
Pedersen, J.S. (2004), *Nye rammer – Offentlig opgaveløsning under og efter strukturreformen*, Copenhagen: Børsens Forlag.
Pheffer, J. and G.R. Salancik (2003), *The External Control of Organizations – A Resource Dependence Perspective*, Stanford, CA: Stanford Business Classics.
Powell, W.W. and P.J. DiMaggio (1991), *The New Institutionalism in Organizational Analysis*, Chicago, IL: The University of Chicago Press.
Scott, W.R. (2001), *Institutions and Organizations*, Thousands Oaks, CA: Sage Publications.
Suchman, M.C. (1995), "Managing legitimacy: strategic and institutional approaches," *Academy of Management Review*, **20** (3), 571–610.
Weber, M. (1976), *Wirtschaft und Gesellschaft: Grundriss der verstehenden Soziologie*, Tubingen: J.C.B. Mohr.
Wenger, E. (1998), *Communities of Practice: Learning, Meaning and Identity*, Cambridge: Cambridge University Press.

PART 3

Positioning

9. The interaction between public science and industry, and the role of the Øresund Science Region's platform organization

Povl A. Hansen and Göran Serin

INTRODUCTION

In the last two decades, the central focus of industrial and scientific policy has been on transferring public scientific research to industry. This has been based on a criticism of the lack of effectiveness of previous policies in this field. Two political instruments have been used, increasing public scientific R&D in general, and developing mechanisms for transferring public science.

The relation between public science and industry has often been seen as "mechanical" in that it has been regarded as a simple transfer of public science to industry. The view reduces public science institutions to transfer institutions and industry to recipients of public knowledge and technology.

This chapter argues that this approach is somewhat simplistic and must be replaced by a more caring approach respecting both the specific characters of both public science and industry, but also paying attention to the specific innovation patterns of different industries and, consequently, their different relations to public science. Only by employing this approach and recognizing these differences is it possible, in the long run, for industry to maximize its benefits from its interaction with public science and maintain the dynamics of both forms of research.

In this connection, a discussion of the character of the interaction between public science and industry has developed. It concerns the impact of public science on different types of industries as well as the importance of diverse types of patterns of interaction that are related to specific sciences and industries. This in turn implies that there are differences in the form of interaction between industry and public science as exemplified by the forms of research co-operation, licensing, scientific publications, conferences, informal meetings and so on. (Mansfield and Lee, 1996;

Cohen *et al.*, 2002; Laursen and Salter, 2004). The dividing line between what academics do and corporate scientists and engineers do is also blurred (Nelson, 1992). This chapter contributes to this discussion by analyzing the patterns and the characteristics of these blurred relations.

Two forms of relations have been identified in the discussion, and have been captured in three models, that is the linear or science push model (Schumpeter, 1934; Bush, 1945) and the demand pull model (Schmookler, 1966). An interactive model combining both of these has also been developed (Gibbons and Johnston, 1975; Kline and Rosenberg, 1986; Nelson, 1990; von Hippel, 1988). The first two models identify the driving forces behind the development of innovation, taking their point of departure in historical facts, while the third model tries to analyse the dynamics of the interaction between science and industry in which feedback processes are seen as central to the developments which take place. The demand-pull and the science-push theories have been subject to strong criticism for being too simple (Mowery and Rosenberg, 1979; Scherer, 1982; Katz and Philips, 1982). This chapter will try to develop and go beyond this rather simplistic characterization of the relation between public science and industry.

First, we stress that there is a difference between the way research is conducted in public science and how it is conducted in industry, and second we discuss the character of science in diverse industries, how it is conducted and what its patterns of interaction with other scientific resources are, and its relation to the firm's innovation processes (Nelson and Winter, 1977; Dosi, 1982). Research in this area often takes its point of departure in specific industries such as biotechnology or the ICT industry (Swann and Prevezer, 1996; Zucker *et al.*, 2002; Owen-Smith *et al.*, 2002; Lim, 2004; Gittelman, 2006).

But there are basic differences in the way innovation processes take place in industry. Firms are very unlike in the way they use knowledge and innovation. This must be recognized in a more caring approach to understanding public science – industry interaction both in the research policy discussion and in the formulation of concrete forms of cooperation.

The most common and important form for innovation in medium and low technology industries is buying and transferring technology and implementing it in the firm. This often takes place through suppliers, between co-operation partners, or as part of an international combine. Innovations in firms often take their point of departure in the existing use of technology. These types of firms have difficulties in absorbing knowledge which is not in the form of a readily usable technology. The innovation processes are addressed toward how technology can be combined and used optimally. Traditionally, universities play a very small role in this connection.

COOPERATION WITHIN R&D

R&D cooperation is the form of innovation most commonly referred to when discussing the transfer of knowledge, this includes public research institutions. External research cooperation also plays a role, especially in the high tech sector, and in firms which have the resources necessary to be part of external R&D co-operation. This applies especially to firms in the pharmaceutical and biotechnological sector, but also to other firms which possess R&D competencies at a scientific level. This can, for example, also be firms in low technology sectors where a firm possesses R&D capacities which make it possible for it to absorb knowledge which does not directly have a material technological form. American research shows that only approximately 10 percent of innovations are based on academic knowledge (Mansfield and Lee, 1996). Even in industries dependent on basic science, only about 50 percent of the firms state that universities and public research institutions are important for their innovations (Laursen and Salter, 2004).

The most common form of co-operation is firms which engage in co-operation with customers as their most important innovation partner and source of inspiration. This is especially the case in medium and low technology industries where co-operation with customers is decisive for their innovation processes. These customers can be professionals who use the product as part of other products. In some cases, the customers themselves take an active part in the innovation processes and they also apply for patents. Cooperation with customers is also decisive in the product's adaptation phase, when it is changed to meet user needs, such cooperation is necessary for the innovation's success. Customer related innovation is the dominant form of innovation, and is crucial, in the service sector, where the service provided is often closely related to the customer's own activity or directly part of it.

Public science on the other hand, is completely different from industrial science both concerning the purpose of research and its organization. The purpose of public science is to solve research issues developed and discussed in the scientific community and not by the market. This is the specific dynamic of the scientific process in public science that it is based on well defined norms and crosses specific interests. The results are published and discussions and criticisms take place in public within the scientific community. This is the driving force behind the development of public science, while the driving force in industrial science is the market and the challenges it is confronted with in that context.

The purpose of this chapter is to understand the dynamics of the scientific processes in the two sectors, and how it is possible for them to interact and co-operate without giving up the dynamics of their own scientific processes and subordinating themselves to the other sector. Investigating the specific forms of interaction and co-operation between the two sectors is a central part of our analysis.

This analysis presented here examines the experience of the Øresund Science Region's platform strategy in relation to establishing knowledge interaction between public science and industry.[1] Its focus will be on the possibilities of developing an organization that is caring when taking into consideration the specific character of public science and industrial research, and the specific character of the innovation processes in different industries and their specific relations to public science.

The research issues presented above will be discussed as follows. First, a definition and discussion of the fundamental character of the interaction between public science and industry will be provided. Central issues in relation to public science – industry linkages will be defined and discussed in an international perspective. Central to the discussion will be the character and the channels of interaction and the possibilities for, and barriers to, interaction.

Taking our point of departure in these central issues we discuss whether the platforms can transcend the role of being knowledge transfer institutions to becoming caring institutions. This means that it is an arena for knowledge interaction between public science and industry where mutual respect is shown for the different character of public science and industrial research. In this connection, the specific character of the platform organizations, in relation to traditional science transfer institutions such as science parks, patents and commercialization institutions, will also be discussed.

The empirical analysis starts with a presentation of the specific organizational structure of the Øresund Science Region and is followed by an interview based analysis of the strategies employed to establish interaction between public science and industry in the different platforms of the Øresund Science Region. The conclusion includes an evaluation of the platform organizations' contribution to the interaction between public science and industry.

THE CHARACTER OF THE INTERACTION BETWEEN PUBLIC SCIENCE AND INDUSTRIAL INNOVATION

In the international discussion about public science-industry linkages, it is possible to identify some issues which are central for analysing the character of the relationships. These will be presented in the following, and later discussed in relation to the platform organization of the Øresund Science Region.

The Contribution of Public Science and the Vertical Chain in Technology Development

Many studies show that public science is not the major contributor to technological development in an industry. Instead, technological development is sourced upstream or downstream in the firm's vertical chain, namely to the firm's suppliers, users or customers (Klevorick *et al.*, 1995). This finding is also supported by a survey carried out by Cohen *et al.* (2002) of 1267 American manufacturing firms which have R&D departments. This survey showed that customers and the firms' own manufacturing operations were among the most important sources of new R&D projects. Of the firms, 90 percent stated that customers were most important and 74 percent their own manufacturing operations. Only 32 percent stated that public research was most important. However, public research was comparable with other sources of knowledge outside the vertical chain such as competitors, consultants/contracted R&D firms and joint or cooperative ventures. It even outscored consultants/contracted R&D as a source of knowledge for both suggesting new R&D projects and contributing to project completion. In a UK survey business units were asked to indicate the importance of different sources of knowledge for innovation. This survey gave internal factors the highest placing followed by suppliers of equipment, materials and components. Customers were given a third place. The number of firms which drew on universities in their innovation activities was only modest (Laursen and Salter, 2004).

Although, interaction within the vertical chain clearly dominates research and technological development, the studies above point to a broader pattern of interaction in which institutions outside the vertical chain are also of importance. This raises the issue of how it is possible to establish institutions or organizations which can incorporate and support this broader pattern of interaction within R&D and technological development. This is an issue we will return to when discussing the Øresund Science Region.

Transfer Contra-interaction in Public Science – Industry Linkages

As previously stated, the interactions between public science and industry are often reduced to simply being an issue of science or technology transfer from public science institutions. This is a consequence of the absence of an interaction perspective on knowledge production, and of the inability to see the different characteristics of the production processes and the implications they have for specific public science-industry linkages. Often this understanding of the transfer of science is further reduced to just incorporating the commercialization of science in the form of patenting, licensing, incubators and the establishment of spin-off firms from universities. Many studies have been undertaken using this perspective (Mowery and Shane, 2002; Zucker *et al.*, 2002; Colyvas *et al.*, 2002; Locket *et al.*, 2005; Markman *et al.*, 2005; Rothaermel and Thursby, 2005). This view is also reflected in concrete policy measures where the focus has mainly been on establishing laws and regulations concerning intellectual property rights that regulate the relations between public science institutions, the individual academic researcher and the enterprise, examples being the Bayh-Dole Act in the US and establishing transfer institutions in universities (Cohen *et al.*, 2002).

In this chapter, by contrast, it is argued that it is important to have a broader understanding, concentrating not just on the transfer of public science but on the interaction which is not just confined to commercialization in a narrow sense. This means that institution building must be broader than just establishing transferring institutions. It being broader makes possible and facilitates a wider range of university-industry interactions.

The Role of Public Science as a Frontier Science or Pool of Knowledge

A central issue in the analysis of public science – industry linkages has been how public scientific knowledge interacts with the industry's knowledge, not least the question of which type of knowledge is most important in this process. A common and traditional view has been that recent science is more valuable to industry than older science, a view which is in line with the linear model. However, during the last decades, an increasing amount of research has pointed to the fact that science as a direct source of new technological ideas are the rarest and therefore the most dramatic cases (Brooks, 1994). The exceptions are biotechnology and to a certain extent pharmaceuticals. The skill-base gathered from most disciplines is more significant than actual research results for developing new technologies. A study carried out in the US showed that only in the disciplines of biology and the medical sciences were the contributions of the skill base and the research results comparable (Pavitt, 1991; Brooks, 1994).

Research on science-industry linkages points to scientific knowledge being a "problem solving pool" which industry can tap when it is confronted with a specific problem which it wishes to solve, rather than being a source of new discoveries. It is precisely this fact which brings out the importance of mutual interaction between the two sides in adapting scientific knowledge to the problems of industry. A large investigation of university-industry linkages in the United States also supports this view: "Overall, university-based research in a field is reported as much less important to recent technological advance than is the overall body of science in the field" (Klevorick *et al.*, 1995). The authors therefore, assert that old science may be as useful as more recent science. It provides a pool of theory, data, techniques and general problem-solving capabilities which influence industrial scientists and engineers. In their investigation of public research's influence on industrial R&D, Cohen *et al.* (2002) point out that "public research plays a slightly more important role overall as a knowledge source for R&D project completion than for initiation."[2] Their findings also suggest that in industrial R&D, personnel contacts to academics, search literature, or forms of cooperative ventures with public research institutions, more commonly address particular needs than they generate new ideas for projects. These two investigations are also in line with an older investigation of the role of scientific knowledge for product innovation in British industry where the authors stated that "the impact of the information transferred was not in the provision of basic ideas for the innovation. Rather scientists tended to respond to previously posed problems . . . Hence, while it is true to say that scientists in universities played little or no role in initiating technological innovation, they frequently made contribution to the resolution of technical problems."[3] This again focuses on the importance of mutual interaction, where industry identifies the relevant problem and interacts with public science to find a possible solution within the existing pool of scientific knowledge.

The fact that it is not primarily new knowledge or front research that is transferred to industry, once again points to the importance of developing organizations to increase the interaction between public science and industry making it possible to integrate the knowledge of the two spheres in a problem solving process. This should be a two way channel of information transfer, bringing both "knowledge pools" together. This chapter will try to show how the platform organizations of the Øresund Science Region are an attempt to establish such a forum for the integration of mutual knowledge.

A central issue in this connection concerns which channels are being used for this interaction between public science and industry. These channels will in turn have implications for how the interaction is built up in organizational terms. In line with that this interaction between industry and public science does not primarily take its point of departure in new discoveries but in the existing pool of public science (Gibbons and Johnston, 1974; Pavitt, 1991; Brooks, 1994; Klevorick *et al.*, 1995; Cohen *et al.*, 2002), it is not surprising that joint ventures and contract research are not the principal channels for interaction between industrial and public science. Cohen *et al.*,'s investigation of the sources of innovation in the American manufacturing sector showed that publications and reports were the dominant channel for information exchange.[4]

Many studies also stress the importance of informal contacts (Crane, 1972; Meyer-Krahmer and Schmoch, 1998; Cohen *et al.*, 2002). In this connection, Rappa and Debackere speak of the significance of technological communities, where academics and industrial researchers form networks of informal contacts (Rappa and Debackere, 1992; Debackere and Rappa, 1994). Both Cohen *et al.*'s study and a German study undertaken by Meyer-Krahmer and Schmoch showed that informal contacts rank number two in importance as information sources for industry and public institutions. Conferences and meetings also rank high in the two studies.

The importance of scientific publishing, informal contacts and meetings and conferences for public science and industry interaction shows the value of open science, and it also shows that the key part of the interaction takes place outside the market.

This fact also points to the significance of taking care and respecting the specific character of public science, namely taking a long term view, being non-profit making, and being internationally oriented toward the free exchange of knowledge in the international scientific community. This places specific demands on the institutions where public and industrial sciences meet and interact, where the character of public science must be preserved, and not be subordinated to the immediate claims of industry. This lack of subordination will also be to the long time advantage of industrial science.

The Importance of Structural Factors for Public Science and Industry Interaction

A central issue in analysing the interaction between industry and public science is the role of structural factors. A common opinion has been that structural factors such as industrial environment, firm size, and R&D intensity play a decisive role in shaping these relations (Klevorick *et al.*, 1995; Cohen *et al.*, 2002). In particular, the firm's absorptive capacities, which are closely related to firm size and investment in R&D, have been considered crucial. Other research (Laursen and Salter, 2004) has also drawn attention to the importance of individual firms' innovation strategies as a condition for interaction with university research, though it does not suggest neglecting the importance of structural factors. This view implies that firms involved in low technology industries, which have more open search strategies, can also be open to interaction with public science

(Chesborough, 2003). However, developing such strategies requires establishing institutions which make interaction with public science possible, especially for firms which normally do not possess such networks with public institutions.

The Triple Helix – a New Way of Establishing Interaction Between Public Science and Industry?

During the last decade public science and industry interaction has to an increasing extent been seen as an instrument for economic growth and the restructuring of the industrial structure by authorities, especially at the regional level. In this connection, the triple helix model has been developed. The model depicts an inter-institutional network representing academia, industry and government (Etzkowitz and Leydesdorff, 1997; Etzkowitz *et al.*, 2000; Etzkowitz, 2002). According to this approach, universities should acquire a third mission in addition to research and education, namely that of being a driving force in responding to a logic of economic development stemming from new interaction networks of academia, industry and government. In the knowledge society, the university, according to this view, is both a cost effective and creative inventor and transfer agent of both knowledge and technology. The institutional configuration of academia, industry and government should displace the military as a leading actor in promoting innovation. Etzkowitz's stage model concentrates on the fact that people from different organizational backgrounds are brought together for the purpose of generating new institutional arrangements, strategies and ideas. The triple helix model also stresses that the relationships between the three helixes are interactive, embedded in projects and in shared values (Etzkowitz, 2002). The triple helix is not expected to be stable. It establishes new networks between academia, industry and government which reshape the underlying institutional arrangements. In this way, layers of communications, networks and organizations among the helixes emerge, while the differences in the underlying institutions are maintained. (Etzkowitz and Leydesdorff, 2000). This process strongly resembles the establishment of the Øresund Science Region which we will analyse later.

However, in terms of concrete research, this approach concentrates on what Etzkowitz (2002) terms the creation of innovation space at the expense of the creation of consensus space. This implies a stronger focus on processes connecting financial and business instruments with new academic ventures and a weaker focus on the process of generating new ideas and strategies by linking the three spheres of the helixes.[5] This in turn results in a relatively traditional research agenda where the linkages between universities and industry are often reduced to knowledge transfers from universities, and the commercialization of knowledge in the form of, for example, spin-offs, patenting and the provision of venture capital by industry.[6] Although there is a pronounced ambition to develop an inter-active model, empirical research has not focused on how this mutual inter-action between the three helixes within science and research develops within the space of the consensus.

We can conclude that a common feature of the issues discussed above is that the linkages between public science and industry have been discussed thoroughly in terms of public science contributing to the development of innovation and technology in industry. This has implied that public science – industry linkages have often been reduced to a one way transfer problem. At the same time, the international research on those linkages dis-cussed above points to problem solving as being their characteristic feature. Problem solving draws our attention to the mutual interaction between public science and industry in which care for the specific characteristics and goals of public and private research is of utmost importance. In this view of innovation, the two forms of research make their own specific contribu-tions. This in turn raises the issue of how this interaction and collaboration should be organized in order to draw on the strengths of both types of research and thereby gain maximum results in both the short and long terms, while at the same time respecting the specific character of public and industrial research. In the following, this issue will be analysed by examining the platform organization of the Øresund Science Region. The topics discussed above will, therefore, be studied empirically employing an interaction perspective. They will be crystallized in the following research questions:

- Have the platform strategies contributed to the interaction between public science and industry or are the contacts reduced to the trans-fer of public science? In what way is the organization constructed to ensure mutual interaction between public science institutions and industry?
- What type of knowledge has been the subject of interaction between public science and industry? Does this collaboration concern new research findings or is it characterized by interaction with the exist-ing pool of knowledge in a problem solving process. Is it possible to identify common tendencies between the different platforms or is the type of knowledge dependent on the specific platform technologies?
- What channels have been important for developing the interaction between public science and industry, such as research projects, con-ferences and informal contacts? Do these channels of interaction differ depending on the character of the platform technologies (for

example the importance of common research projects, conferences and meetings)?

- Has the interaction between public science and industry been "structurally" restricted to R&D intensive or large firms or have the platforms been able to reach other firms? Has the interaction been dependent on the character of the platform technologies or has it been the result of deliberate platform strategies?
- Has the Øresund Science Region been able to deepen the interaction within the triple helix to include knowledge and technology interaction between academia, industry and government? What is the specific contribution of regional authorities to innovation and technological development in the region?

THE ØRESUND SCIENCE REGION AND THE INTERACTION BETWEEN PUBLIC SCIENCE AND INDUSTRY

A special characteristic of the organization, the Øresund Science Region, which we will discuss in detail here, is that it was established within a broader cooperation between Denmark and Sweden. In this cooperation, a number of organizations have been created. Characteristic of them is that they have been set up "outside" the public institutional structure, but in cooperation with existing institutions on both sides of the border. The problem common to these new institutions is their lack of power and formal authority when pursuing the integration process. On the other hand, this also presents strong possibilities for trying out new methods and organizations in cross-border co-operation.

They are usually informal organizations which primarily argue and lobby for changes in laws and regulations but do not have the authority to effect change. For the most part, they are organizations working to establish public or private sector networks, identifying opportunities for cooperation and raising the necessary funds for the integration process. Their other valuable function is to draw the attention of the relevant regional and national authorities to barriers to the integration and cooperation processes before, and as, they emerge.

Here the Øresund Science Region is an important organization, its purpose being to strengthen research and cooperation between research institutions and industry on both sides of Øresund. The organization is a sub-organization of the Øresund University, which is a co-operation between 14 universities in the Øresund Region. The important role of the university is primarily based on the regional strategy for closer and stronger

The **Øresund Science Region,** is an umbrella organization for a number of "platforms" or networking agencies which have been established for developing cooperation across the Øresund between players and institutions in the fields of innovation, technological development and knowledge transfer. It was established in cooperation between industry, public authorities and universities.

The organization structure of the Øresund Science Region:

Board and secretariat (Representatives from industries, regional authorities and universities)

↓

Øresund IT Academy
Board (Representatives from industry, regional authorities and universities).

Director and secretariat.

| Research field | Research field |

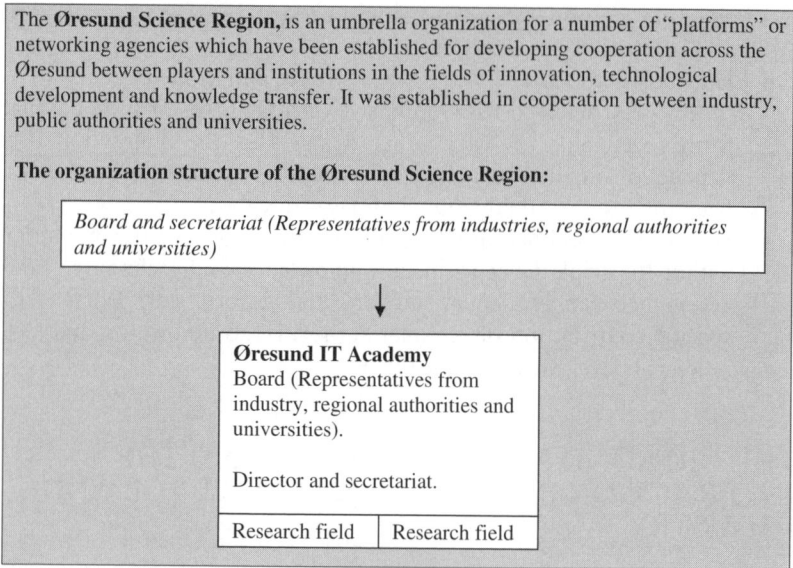

Figure 9.1   The Øresund Science Region

integration between science institutions and industries in the region. The Øresund Science Region is an alliance of seven regional research and innovation platforms established by universities, industries and local authorities on both sides of the sound.

The organization structure consists of the following platforms: Nano Øresund, Diginet Øresund, the Medicon Valley Academy, the Øresund IT Academy, the Øresund Food Network, the Øresund Environment Academy, and Øresund Logistics. The total staff of the Øresund Science Region amounts to approximately 40 persons.

The seven regional platforms are organized relatively autonomously with their own boards and secretariats, and with their own managing directors leading the organization.

The main objective of these network institutions is to establish cooperation between science institutions and local business in order to strengthen both public institutional research, the cooperation between institutional and industrial research, and R&D co-operation between firms on both sides of the border. These organizations can both be seen from the point of view of local universities and industries wishing to establish a more research based regional development, but can also be seen as an answer to the general problem found in both Denmark and Sweden of developing a more science based industry.

THE ROLE OF THE PLATFORM ORGANIZATION OF THE ØRESUND SCIENCE REGION IN DEVELOPING THE INTERACTION BETWEEN PUBLIC SCIENCE AND INDUSTRY

The central issue of this section is to analyse the role of the platform organization of the Øresund Science Region and its possibilities to go beyond the traditional role of a transfer institution and becoming a caring knowledge institution by establishing interaction between public science and industry. We take our point of departure in understanding the specific dynamics of the scientific processes in the two sectors and the specific conditions and resources of the different industries for establishing such interaction. The following section is an empirical analysis of the strategies of the different platforms to establish such an interaction.

Public Science and Industry Relations as an Interactive Process

Following the linear theory, the discussion of public science and industry linkages has been concerned with how industry can make use of knowledge developed within public science. This has been seen as a relatively uncomplicated process, at least because one only has to consider this process as a one way flow of knowledge and technology from public science institutions to industry. This has in turn implied that the focus has only been on facilitating this transfer process. This has resulted in a concentration on the character of public research making the research immediately more attractive to industry (Globaliseringsrådet, 2006). This implies that public research should to a greater extent respond to the needs and interests of industry. The concept the "entrepreneurial university" is an expression of this. In the short run, this can be interpreted as that public science should be governed more by profit and have a market orientation. Another consequence of the "transfer" view is that the absorptive capabilities of the receiver of this knowledge transfers become critical. This in turn implies restrictions on the receiver to only include firms with large R&D intensities like, for example, biotechnology and ICT firms.

But this focus on the transfer of knowledge has also resulted in a focus on institution building to facilitate these transfers. Central to this has been the concentration on innovation speed, that is the speed of transferring university technology to the market (Markman *et al.*, 2005). During the last two decades, a wide range of institutional arrangements for the commercialization of public science has been developed such as science parks and university technology transfer offices.

These traditional institutional arrangements for knowledge and technology transfer are of course also present in the Øresund Region. What is special in the Øresund Region is the deliberate attempt by the Øresund Science Region to complement these one way transfer institutions with a more interactive network organization.

The Øresund Science Region's platform organization is a deliberate attempt to build up a network organization. As a network organization, the interaction between the different actors of the network is a crucial part of the strategy, and this is expressed, as shown above, in the make up of the board of the Øresund Science Region where representatives from industry, academia and local authorities are present, and in the boards of the different platforms. This constitution of the board ensures a prerequisite for the interaction. The organization of platforms around specific industries or sectors also allows this interaction to take specific forms in accordance with the special characteristics of the platforms' technology. As pointed out before, the Øresund Science Region with its concentration on overlapping networks can be seen as a concrete expression of a triple helix model, although the structure of the underlying institutions in the form of universities, industry and government remains unchanged. In the following, the strategies of the different platforms to establish this interaction between public science and industry will be discussed starting with the central issues presented above.[7]

What Type of Knowledge -- New Research or Tapping the Pool of Knowledge?

The different platforms regard the mediation of research contacts between firms and between firms and public science on both sides of the border as their prime mission. It is expressed clearly by the Managing Director of Medicon Valley Academy: "Our role is to find out where the human resources can be found. Where do we find knowledge and competences within different fields of research? This is done within a defined field" (Managing Director Stig Jørgensen MVA).

In this way, we can define the platforms as organizations which facilitate the ability of different actors in the region to tap the "pools of knowledge" that already exist in the region both in public institutions and in firms. A central function for the platform secretariats is, therefore, to map the region in terms of the competences and knowledge within the field. This is a function which an individual firm seldom has the resources to undertake. Now the firms can apply to a single institution in the region. In this way the search costs for the firms can be reduced considerably. Another advantage, besides its resources, is that the platform is not biased in relation to specific interests and is non-commercial in nature. This implies that the platforms are considered as being trustful, which again reduces transaction costs. All this increases the possibilities to develop and establish knowledge networks both with public science institutions and private firms.

Some platforms also focus on branding the platforms internationally and establishing contacts with similar organizations in other science and knowledge regions. This has been part of the strategy of Medicon Valley Academy which has established contacts with cluster organizations for example in Cambridge, Kobe and Boston. In this way, the platform can contribute to tapping an international "pool of knowledge" and again resulting in lower search and transaction costs.

Which Channels of Interaction are Being Used by the Platforms?

The channels of interaction between public science and industry differ greatly between the diverse platforms of the Øresund Science Region dependent on the character of the platform technology.

Common to all platforms, however, is an attempt to establish interaction and contacts between and within the three spheres of the triple helix. The platforms state the importance of showing respect for the different character and interests of the different actors when establishing these contacts (Managing Director Peter Höjerback Øresund IT Academy (ØIA) and Knowledge Director Henriette Moos Diginet Øresund (DØ)).

An important part of the activities of all platforms is to arrange conferences, seminars and workshops. These can be of various types, more popular ones where a broad range of actors are invited, or narrower research networks where only specific researchers and firms participate. The platforms with their secretariats also have the resources and overview which makes it possible for them to identify topics and themes which can be relevant for different actors in the region. In this way, these meetings are important instruments for establishing specific networks between different actors within public science, industry and authorities.

As was found in the international discussion, informal contacts play a significant role in scientific contacts between public science and industry. A lot of spontaneous contacts arise out of conferences and meetings in the platforms. The problem is that people seldom have the time and the resources to develop these contacts. Here the secretariat has the resources to develop such informal, spontaneous contacts into more structured relationships.

There are clear differences between the platforms concerning their view of their central tasks. The more high tech platforms like Øresund IT Academy, Medicon Valley Academy and Nano Øresund see their prime task as mediating contacts between firms, between firms and researchers and between researchers. In the more low technology platforms like the

Øresund Food Network (ØFN), Øresund Logistics (ØL) and the Øresund
Environment Academy (ØEA) the platform plays a much more active role
in the mediation of contacts. This difference in approach comes to expression in the varying importance of research projects in the two types of platforms, where the platforms take a much more active part in establishing
research projects in the latter than in the former. We can see diverse patterns of interaction when we study the different platforms.

The Medicon Valley Academy is clearly oriented toward the commercialization of research. In this respect, the platform resembles traditional
transfer institutions more than the other platforms. It very seldom has an
active role in formulating research projects involving different actors. This
does not mean that the universities are not part of the firms' normal
research, but that the firms can manage many of its research problems
alone. When competencies are missing, the firms contact researchers they
know, or scan the relevant literature to search for researchers with the
necessary competencies.

It is characteristic of the IT field that the firms often have large R&D
resources of their own (Peter Höjerbach ØIA). A lot of research takes place
within the firm's R&D department and there is a basic difference between
public and private research as pointed out by the Managing Director of
Øresund IT Academy:

> The agenda is different. The public researchers are there to develop new
> knowledge – look at things in a new perspective. The firms must produce
> profits. This is the core problem. This has the effect that the firms must
> commercialize their results quickly. The time horizons between firms and
> public science are very different. The research in the IT firms is much applied.
> (Peter Höjerbach ØIA)

As a result of the platform activities, it is possible for firms to understand
that there must be a research component, and public researchers to understand that the results must be applicable if a co-operation between public
research and industry is to be fruitful for both. Just as was the case with
Medicon Valley, the platform plays a very small role in initiating research
projects. The role is more to mediate contacts.

Another high technology platform is Nano Øresund (NØ). At the
time of writing the platform had played no role in developing research
projects. These are initiated by the university institutions themselves. The
researchers have their own personal networks but the large firms are to
some extent part of these networks. The platform has tried to map the
needs of different firms and to identify which firms can be of interest in
relation to cooperation within science and technology development. Some
large firms have their own R&D resources and have not been interested in

cooperation while the smaller R&D intensive firms on the other hand have shown interest in cooperation within the platform. Because the technology is still unknown to many firms, the platform has used a lot of time presenting the possibilities of nano technology for firms. They have also travelled around to different institutions related to the platform telling them about their capabilities (Managing Director Anne Line Mikkelsen NØ).

The Diginet Øresund platform organizes the experience industry. Traditionally, there have been weak ties between product development and public research in this industry. "The field was defined by some of the industry's entrepreneurs and not by the researchers" (Knowledge Manager Henriette Moos DØ). In the beginning public research was not particularly interested in co-operation with industry. Instead, their focus was on the use of media or the culture which is related to different media. In this context, research projects are important for establishing interaction between the different actors. Here the platform plays an active part because of the weak industrial tradition to interact with public science on one hand and the low interest of the universities to engage in research projects with industry on the other.

The environmental industry is characterized by being a low technology industry, where there are few private resources for R&D. Alternatively there is a large amount of public research. "This research is addressed to basic issues within the field, while the firms often have a more practical orientation" (Managing Director Jacob Juul ØEA). For these firms there is also a hard economic reality, which influences their ability to participate in research projects. As in the other low technology platforms, Øresund Environment Academy has played an active role in initiating and developing the research projects.

Øresund Logistics is another low technology platform. Public science, together with the public authorities plays an important role in this platform, in contrast to the other platforms. The reason for this is that public authorities have large interests in the development of transport in the form of transport infrastructure, which is very costly (Project Manager Patrik Rydén ØL). In this platform, the research projects are initiated by public science and public authorities. The problems for private firms in participating are related to resources, personal and the time horizons of the projects. This implies that only large firms, within packaging, transport and IT, participate in the large projects.

Finally, we have Øresund Food Network (ØFN). This platform consists of firms which are R&D based, either large research-based firms or small firms that are spin offs from university research. These firms can be called the A group. On the other hand there is a group of low technology firms with no contact to public research – the C group. Gradually, a group of B firms has been developed with a larger capacity to absorb science (Director of Research Maria Olofsdotter ØFN). A strategic interest of the platform is to come in contact with the C group – an issue which we will return to in the next section. The distance between public research and the R&D based firms is not so great. Both in Denmark and Sweden there is a tradition for cooperation between public science and industry (Director of Research Maria Olofsdotter ØFN).

Also in the Øresund Food Network an important part of the interaction between industry and public science has taken the form of research projects where the platform has played a decisive role.

Thus, we can see that the channels of interaction between industry and public science differ between the platforms especially dependent on the technological characteristics of the platform. A crucial factor in this case is whether the firms of the platform are based on high or low technology. In the case of low technology platforms, research projects play a much more important role. Here the platform also plays a decisive role both as an initiator and organizer of the research projects. This is due to the lack of competencies, absorptive capacity and general view in the firms. These factors which are necessary both to identify common research issues and to establish a structural cooperation are factors which the platform secretariat has. These are on the other hand capabilities which the firms in high technology platforms possess. This implies that the high technology platforms focus more on establishing interaction between the actors of the platform, where the platform functions more as an intermediary between the actors.

Developing New Structures Between Public Science and Industry

As pointed out in the international discussion about public science-industry linkages, a problem has been that the linkages have been structurally restricted to firms which are able to absorb R&D results produced by public science – firms in the high tech sector such as pharmaceuticals, biotechnology and ICT. Besides this structural linkage, we can also add another structural restriction, which consists of public science – industry linkages which traditionally have been developed within specific scientific fields. Although the platforms are established within these fields and primarily establish contacts within them such as IT and biotechnology, the co-existence of the different platforms within the same organization, the Øresund Science Region, also makes it possible for cross-platform contacts. This is also facilitated by the fact that the secretariats of the different platforms are geographically located in the same building. This gives plenty of opportunities for contacts, not the least of an informal nature, between the personnel of the different platforms. These informal contacts have been important for example for developing cross-platform projects. Examples of such cross-platform projects are "Food and IT" and "Borderless Health" the latter between the Medicon Valley Academy and Øresund Food Network. Other examples of cross-platform projects are Bio+IT which was a collaboration between the Medicon Valley Academy and the Øresund IT Academy. These cross-platform projects can be seen as an important way of breaking up traditional industrial and academic trajectories.

But some platforms have also been important in relation to the more classical structural problem of reaching firms with low absorptive capacity of public science that is firms which are traditionally labelled "low tech." The platform Øresund Food Network has deliberately tried to get into contact with low technology firms in the science and technology field by trying to incorporate them in research projects together with larger food firms and public science institutions. In this way, the platform attempts to increase the innovative power of these low technology firms by trying to stimulate research interest and get academics employed in the firms. Here it is also important for the platform to find research themes that can attract low tech firms. In this specific case, the platform formulated a research project around "Food products and IT." Because today IT is also a part of daily life of low tech firms, it can be a way of linking these firms to public research. In this project, firms' suppliers also participate and the low tech firms play an important role because of their user competences. This type of project is also a way for low tech firms to get access to leading researchers (Director of Research Maria Olofsdotter ØFN). This requires that the platform has a strong control over the research project.

But the platform organizations also interact with smaller firms which do not normally have relations to public science. This is done through seminars and meetings which have an agenda that also attracts these firms (Managing Director Jacob Juul ØEA, Director of Research Maria Olofsdotter ØFN). The access to many instruments therefore facilitates the platform's interaction with firms that traditionally have problems with absorbing science.

We can thus see that the platform organization makes it possible to break up traditional structures between scientific disciplines and between public science and industry. But we can also see that the platform organization expresses care, when it makes interaction between sciences possible while respecting the autonomy and the specific character of sciences.

The Role of Triple Helix in the Platform Organization

As the international discussion indicated, the interaction of the triple helix of academia, industry and government has often been restricted to knowledge transfers from academia, industry funding and the government providing optimal institutions for university-industry interaction. The triple helix has to a very small extent concerned interaction between these institutions within the field of science and technology. As becomes evident from the previous sections, the important contribution of the platform organization of the Øresund Science Region is that this interaction primarily takes place within science and technology development. This is not only expressed in the organizational structure where the three helixes are not only represented on the board of the Øresund Science Region, but also on the board of each of the platforms. It is primarily expressed in the practices of the different platforms where the prime mission is to establish meeting places, contacts and cooperation through research collaboration, conferences and seminars.

Especially noteworthy is the role of the regional authorities in science and technology development. These are not only reduced to providers of institutional settings but are also part of the interaction within science and development – not the least because of their user competencies. Within the Øresund Environment Academy, the local authorities, especially within planning and regulation, play an important role in building the projects because they have a close interest in them. But they do not influence the strategic development because they shall not govern the research in the platform (Managing Director Jacob Juul ØEA). There is also a close dialogue with the authorities in areas like Novel and Functional Foods, where regulations are important and where these authorities are very influential in relation to the specific development of the innovation process (Director of Research Maria Olofsdotter ØFN). Beside these functions, the authorities are important for financing the activities of the platforms and not the least to sell the strategies of the platforms to the politicians.

The biggest contribution of the platform organization of the Øresund Science Region has therefore been to deepen the interaction in the triple helix model by focusing on the interaction within the core field, namely science and technology.

CONCLUSION

There are large differences in the way firms absorb and use knowledge and innovation. In the international discussion of public science industry linkage, insufficient consideration has been given to this central issue. The linkage between public science and industry is still primarily regarded as a one way transfer process of public knowledge and technology to industry and where commercialization of public knowledge is of prime interest and the recipients are high tech industries with a large capacity to absorb science.

The platform organization of the Øresund Science Region is an attempt to break up this one way transfer perspective by developing a caring approach. This is done by introducing a truly interactive network perspective within a triple helix model, where the specific network interaction is adopted to the specific technology of the platform. In this type of co-operation, it is also possible to use and maintain the specific characteristics of public science and industry research, while at the same time paying attention to the specific innovation patterns of the individual industries.

There are many strengths but also weaknesses in the way the platforms are organized. The strength is the flexibility and dialogue that characterize the platforms. In this way, respect is shown for the way research is conducted in different sectors – both within the public and private sectors, but also within various environments. The many different instruments that the platforms possess to initiate these interactions from conferences, seminars and workshops to research projects makes it possible for the interaction to take different forms depending both on the specific issues and the participants. This will also ensure a mutual interaction instead of a one way transfer. It has also been shown that the platforms interact within different forms of knowledge dependent on the industries concerned. The large tool box has implied that the platforms can interact with industry independent of firm size. Another advantage is that national and local authorities are part of the platform and hereby involved in the innovation processes and its consequences.

But there are also problems connected to the platform organization. There is limited engagement from individual public researchers. There is also a lack of knowledge of the Øresund Science Region in large parts of the research community. To this can be added formal barriers between Denmark and Sweden (Garlick *et al.*, 2006) and the lack of project money that is not nationally bound but can be used across the border. Another problem is that the institutions have a clear regional perspective, while research and development is addressed to an international research community. These are, however, problems that are now being tackled within the platforms. If these problems can be solved, the platform organization with its flexibility and interaction will be an interesting alternative to the traditional transfer view of public science and industry relations.

NOTES

1. The Øresund Science Region is a cross border institution the purpose of which is to support innovation by establishing co-operation between firms and between firms and public science institutions across the border. This will be discussed in depth later.
2. Cohen *et al.* (2002) p. 7.

3. Gibbons and Johnston (1974) p. 235.
4. If only viewed from the university point of view, a German investigation of university-industry interaction showed that collaborative research with industry was the most important form of interaction, while publication was of minor importance. (Meyer-Krahmer and Schmoch, 1998). From a university point of view this result could hardly be surprising.
5. A good example is the great importance Etzkowitz attaches to invention of the venture capital firm in New England and how it was related to Harvard and MIT (Etzkowitz and Leydesdorff, 2000 and Etzkowitz, 2002).
6. There are many examples of research projects in this tradition. See for example Sapsalis *et al.*, 2006; Shinn and Lamy, 2006; Avnimelech and Teubal, 2006; Mueller, 2006; Feldman and Kelley, 2006; Lawton Smith and Ho, 2006 and Landry *et al.*, 2006.
7. The discussions are based on interviews with central representatives of the different platforms.

REFERENCES

Avnimelech, G. and M. Teubal (2006), "Creating venture capital industries that co-evolve with high tech: insights from an extended industry life cycle perspective on the Israeli experience," *Research Policy*, **35** (10), 1477–98.
Brooks, H. (1994), "The relationship between science and technology," *Research Policy*, **23** (5), 477–86.
Bush, V. (1945), *Science the Endless Frontier: A Report to the President on a Program for Postwar Scientific Research*, Washington, DC: US Government Printing Office.
Chesbrough, H. (2003), *Open Innovation*, Cambridge, MA: Harvard University Press.
Cohen, W.M., R.R. Nelson and J.P. Walsh (2002), "Links and impacts: the influence of public research on industrial R&D," *Management Science*, **48** (1), 1–23.
Colyvas, J., M. Crow, A. Gelijns, R. Mazzoleni, R.R. Nelson, N. Rosenberg and B. Sampat (2002), "How do university inventions get into practice?," *Management Science*, **48** (1), 61–72.
Crane, D. (1972), *Invisible Colleges*, Chicago, IL: Chicago University Press.
Debackere, K. and M.A. Rappa (1994), "Technological communities and the diffusion of knowledge: a replication and validation," *R&D Management*, **24** (4), 355–71.
Dosi, G. (1982), "Technological paradigms and technological trajectories – a suggested interpretation of the determinants and directions of technical change," *Research Policy*, **11** (3), 147–62.
Etzkowitz, H. (2002), "The Triple Helix of University-Industry-Government. Implication for Policy and Evaluation," Stockholm: Institutet för studier av utbildning och forskning, working paper, 2002:11.
Etzkowitz, H. and L. Leydesdorff (1997), *Universities and the Global Knowledge Economy: A Triple Helix of University-Industry-Government Relations*, London: Pinter.
Etzkowitz, H. and L. Leydesdorff (2000), "The dynamics of innovation: from National Systems and 'mode 2' to a triple helix of university-industry-government relations," *Research Policy*, **29** (2), 109–23.

Etzkowitz, H., A. Webster, Chr. Gebhardt and B.R.C. Terra (2000), "The future of the university and the university of the future: evolution of ivory tower to entrepreneurial paradigm," *Research Policy*, **29** (2), 313–30.

Feldman, M. and M. Kelley (2006), "The *ex ante* assessment of knowledge spillovers: government R&D policy, economic incentives and private firm behaviour," *Research Policy*, **35** (10), 1509–21.

Garlick, S., P. Kresl and P. Vaessen (2006), *Supporting the Contribution of Higher Education Institutions to Regional Development*, peer review report on the Øresund Science Region, a cross border partnership between Denmark and Sweden, Paris: OECD.

Gibbons, M. and R. Johnston (1974), "The roles of science in technological innovation," *Research Policy*, **3** (3), 220–42.

Gittelman, M. (2006), "National institutions, public-private knowledge flows, and innovation performance: a comparative study of the biotechnology industry in the US and France," *Research Policy*, **35** (7), 1052–68.

Globaliseringsrådet (2006), *Fremgang, fornyelse og tryghed*, Copenhagen: Statsminis-teriet.

Katz, B.G. and A. Phillips (1982), "Government, technological opportunities and the emergence of the computer industry," in H. Giersch (ed.), *Emerging Technologies*, Tübingen: J.C.B. Mohr, 419–66.

Klevorick, A.K., R.C. Levin, R.R. Nelson and S.G. Winter (1995), "On the sources and significance of industry differences in technological opportunities," *Research Policy*, **24** (2), 185–205.

Kline, S.J. and N. Rosenberg (1986), "An overview of innovation," in R. Landau and N. Rosenberg (eds), *The Positive Sum Strategy*, Washington, DC, National Academy Press.

Landry, R., N. Amara and I. Rherrad (2006), "Why are some universities more likely to create spin-offs than others? Evidence from Canadian universities," *Research Policy*, **35** (10), 1599–615.

Laursen, K. and A. Salter (2004), "Searching high and low; what types of firms use universities as a source of innovation?" *Research Policy*, **33** (8), 1201–15.

Lawton Smith, H. and K. Ho (2006), "Measuring the performance of Oxford University, Oxford Brookes University and the government laboratories, spin-off companies," *Research Policy*, **35** (10), 1554–68.

Lim, K. (2004), "The relation between research and innovation in the semiconductor and pharmaceutical industries (1981–1997)," *Research Policy*, **33** (2), 287–321.

Lockett, A., D. Siegel, M. Wright and M.D. Ensley (2005), "The creation of spin-off firms at public research institutions: managerial and policy implications," *Research Policy*, **34** (7), 981–93.

Mansfield, E. and J.Y. Lee (1996), "The modern university: contributor to industrial innovation and recipient of industrial R&D support," *Research Policy*, **25** (7), 1047–58.

Markman, G.D., P.T. Gianiodis, P.H. Phan and D.B. Balkin (2005), "Innovation speed: transferring university technology to market," *Research Policy*, **34** (7), 1058–75.

Meyer-Krahmer, F. and U. Schmoch (1998), "Science-based technologies: university-industry interactions in four fields," *Research Policy*, **27** (8), 835–51.

Mowery, D.C. and N. Rosenberg (1979), "The influence of market demand upon innovation: a critical review of some recent empirical studies," *Research Policy*, **8** (2), 102–53.

Mowery, D.C. and S. Shane (2002), "Introduction to the special issue on university entrepreneurship and technology transfer," *Management Science*, **48** (1), v–ix.

Mueller, P. (2006), "Exploring knowledge filter: how entrepreneurship and university-industry relationships drive economic growth," *Research Policy*, **35** (10), 1499–508.

Nelson, R.R. (1990), "Capitalism as an engine of progress," *Research Policy*, **19** (3), 193–214.

Nelson, R.R. (1992), "What is 'commercial' and what is 'public' about technology, and what should be?," in N. Rosenberg, R. Landau and D.C. Mowery (eds), *Technology and the Wealth of Nations*, Stanford, CA: Stanford University Press.

Nelson, R.R. and S.G. Winter (1977), "In search of a useful theory of innovation", *Research Policy*, **6** (1), 36–76.

Owen-Smith, J., M. Riccaboni, F. Pammolli and W.W. Powel (2002), "A comparison of US and European university–industry relations in the life sciences," *Management Science*, **48** (1), 24–43.

Pavitt, K. (1991), "What makes basic research economically useful?" *Research Policy*, **20** (2), 109–19.

Rappa, M.A. and K. Debackere (1992), "Technological communities and the diffusion of knowledge," *R&D Management*, **22** (2), 209–20.

Rosenberg, N. (1990), "Why do firms do basic research (with their own money)?" *Research Policy*, **19** (2), 165–74.

Rothaermel, F.T. and M. Thursby (2005), "Incubator firm failure or graduation? The role of university linkages," *Research Policy*, **34** (7), 1076–90.

Sapsalis, E., B.V. de la Potteriea and R. Navon (2006), "Academic versus industry patenting: an in-depth analysis of what determines patent value," *Research Policy*, **35** (10), 1631–45.

Scherer, F.M. (1982), "Inter-industry technology flows in the United-States," *Research Policy*, **11** (4), 227–45.

Schmookler, J. (1966), *Invention and Economic Growth*, Cambridge, MA: Harvard University Press.

Schumpeter, Joseph (1934), *Theory of Economic Development*, Cambridge, MA: Harvard University Press.

Shinn, T. and E. Lamy (2006), "Paths of commercial knowledge: forms and consequences of university-enterprise synergy in scientist-sponsored firms," *Research Policy*, **35** (10), 1465–76.

Swann, P. and M. Prevezer (1996), "A comparison of the dynamics of industrial clustering in computing and biotechnology," *Research Policy*, **25** (7), 1139–57.

von Hippel, E. (1988), *The Sources of Innovation*, New York: Oxford University Press.

Zucker, L.G., M.R. Darby and J.S. Armstrong (2002), "Commercializing knowledge: university science, knowledge capture, and firm performance in biotechnology," *Management Science*, **48** (1), 138–53.

10. The role of a network organization and Internet-based technologies in clusters: the case of Medicon Valley

Ada Scupola and Charles Steinfield

Over the last 20 years there has been much interest in the concept of industrial localization, the tendency for many industries to concentrate specialized activities in particular locations, as a source of innovation and competitive advantage (for example Porter, 1998; Becattini, 1990). A competing view, on the other hand, is offered by researchers who claim that the advent of globalization and ICTs would pose a challenge to local economies (Cairncross, 1997). This chapter addresses these competing notions by investigating the role of a network organization, Medicon Valley Academy, in supporting the development of a Danish biotech cluster, Medicon Valley. Rather than viewing globalization and localization as competing perspectives, the chapter shows the important role that Internet-based technologies have both in supporting coordination and collaboration inside industrial clusters, and in contributing to the globalization of the economy by connecting companies and clusters of companies across different regions of the world.

INTRODUCTION

Innovation and technological change has been an important factor for economic development. In knowledge-based economies innovations are often the output of interactions and collaboration between various actors rather then being the result of a single firm or a single individual (Malmberg and Power, 2006). In the last two decades there has been much interest in the concept of industrial localization, the tendency for many industries to concentrate specialized activities in particular locations as a source of innovation and competitive advantage (for example Becattini, 1990; Porter, 1998; OECD, 2002). Some researchers (for example Cairncross, 1997) have claimed that globalization and the movement toward an information technology-driven economy challenges the significance of location and spatial proximity for business performance and threatens the success of local clusters all over the world (Asheim *et al.*, 2006). Cluster proponents, however, (for example Porter, 1998) remain convinced that globalization and the movement toward an information technology-driven economy increases the importance of location for economic activities. As Porter (1998) states, we are in front of a paradox: on one hand we have a global economy, on the other hand we have the growing importance of local economies constituted of geographic, cultural and institutional proximity. Nevertheless, as pointed out by Steinfield and Scupola (2006) the role of information and communication technologies (ICTs) in the development and maintenance of local industrial clusters has not received much attention in the cluster literature, despite the investments that many governments are making in ICT infrastructures such as for example broadband. Here Porter (1998)'s definition of clusters is adopted according to which "clusters are geographic concentrations of interconnected companies and institutions in a particular field" (Porter, 1998: 78). These institutions can provide specialized training, education, information, research and technical support. They include governmental institutions, universities, standards-setting agencies, think tanks, vocational training providers and trade associations.

The purpose of this chapter is to investigate how Internet-based technologies can contribute to the paradox mentioned by Porter. That is on the one hand ICTs support localized economies such as industrial clusters and on the other hand they contribute to the globalization of the economy by connecting companies and clusters of companies with international stakeholders. To do this, the chapter analyses how a network organization has used Internet-based technologies to increase collaboration and knowledge exchange among actors located within a cluster and between the cluster and other regions of the world. Also the chapter shows how the network organization in question has used Internet-based technologies for increasing cluster visibility and branding. In addition, by referring to the theme of this book which is "Innovation with care," here "care" has the meaning of "taking care of" in the sense of supporting the cluster, establishing a cluster brand, promoting the cluster internationally, promoting innovation in the region and attracting international labor and companies to the region.

The cluster in consideration is Medicon Valley, a biotechnology cluster located in Denmark and Southern Sweden. The network organization is the Medicon Valley Academy (MVA) that, since its establishment in 1996, has had an important role in the support of the cluster both locally and internationally. The analysis is based on empirical material from a larger project on the use of ICTs in clusters conducted by the authors. Specifically, the analysis is mainly based on interviews conducted with two different representatives of Medicon Valley Academy (MVA), information provided at MVA's web site and other documentary material. Quotations from interviews are provided to illustrate some major points and interviewee perceptions.

The chapter is structured as follows. This section has introduced the background and purpose of the study. The next section presents a short review of different forms of localized concentrations of specialized activities. This is followed by a brief overview of models that explain economic development and a literature review of ICTs in clusters. The rest of the chapter provides first some background information about the Medicon Valley cluster and the region where it is located. Then it presents an analysis of the supportive (taking care of) role of Medicon Valley Academy for the cluster and how Internet-based technologies have contributed to this. Finally some concluding remarks are given.

AN OVERVIEW OF THE CONCEPT OF CLUSTER AND RELATED CONCEPTS

Over the past two decades or so there has been much interest in industrial localization from different disciplines: economic geography, business studies and economics (Asheim *et al.*, 2006). The localization of industrial activities takes different forms and different names have been used to describe them. The industrial districts of the Third-Italy were among the first to be studied and have been a source of inspiration for many other forms. Other forms include the new industrial spaces of California (Scott, 1988), local and regional innovation systems (for example Cooke, 1998 and 2001; Lundvall, 1993), local production systems (Crouch *et al.*, 2001), and the concept of the local and regional business cluster (Porter, 1998). However the concept of clusters and industrial districts is not new. In fact, Marshall (1936) already talked about the external economies of industrial districts at the beginning of the century. Marshall conceived of the industrial district as a logical outcome of the process of economic evolution and "saw industrial districts as an alternative mode of industrial organization to the large integrated firm with its internal economies of scale" (Asheim *et al.*, 2006: 6). In an attempt to conceptualize the cluster concept, Asheim *et al.* (2006) distinguish five main theoretical perspectives:

- Italian Neo-Marshallian Industrial Economics: this approach focuses on external economics, inter-firm division of labor, and social capital.

It is characterized by local industrial districts of export-oriented small firms and flexible specialization (for example Piore and Sabel, 1984).

- New Trade Theory and Marshallian Localization Economics: focuses on external economies and increasing returns as bases of trade. It is characterized by geographical agglomeration and localized specialization of industry (for example Krugman, 1998).
- New Endogenous Growth Theory: Focuses on educated labor and R&D as sources of increasing returns. It is characterized by localized technological progress and divergent regional growth (for example Martin and Sunley, 1998; Loasby, 1998).
- Economics of firm strategy and Marshallian localization economics: focuses on external economies and competitive rivalry. It is characterized by local clustering as a driver of productivity and competitiveness (for example Porter, 1998).
- Neo-Schumpeterian and Evolutionary Economics: focuses on institutions, innovation and learning. It is characterized by local entrepreneurial milieux, learning regions and regional path dependence (for example Cooke, 1998 and 2001; Lundvall, 1993).

However, whatever the form of industrial localization or its theoretical explanation, one of the main expectations is that it produces innovation and competitiveness through interaction among a number of actors that are located within a well specified geographical region. This can happen in several ways. The ones most commonly mentioned in the literature include face-to-face interactions, common language and trust among the different actors (Malmberg and Power, 2006; Malmberg and Maskell, 2002).

In the last decade the cluster concept, mainly inspired by the work of Porter (1990; 1998), has become the standard term for localized specialization of economic activities, and it has been used not only as an analytical tool, but also as a policy tool (Asheim *et al.*, 2006). According to Rosenfeld (2001) Denmark was the first country to use policy levers to establish networks and clusters of companies. Denmark therefore has been the first to translate the concept of industrial districts and inter-firm cooperation, formerly associated primarily with the cultural and business environment of Italy, into a more universal practice (Rosenfeld, 2001). The Danish example has since been followed by many other countries such as Spain and the USA. However, as already stated, most of the literature on clusters and economic agglomerations have not dealt with the question of how ICTs can impact clusters and the relationships and interactions among the cluster members that seems to be so crucial to their very existence. A review of the literature addressing this is therefore provided in the next section.

ICTs AND CLUSTERS

Even though the literature on clusters and relative types of localized economies has been notably increasing over the last two decades (see Maskell and Kabir, 2006 for an overview of the number of publications in scholarly journals with the term cluster or its synonyms), the literature on the role of ICTs in clusters is still limited. Leamer and Storper (2001) point out that the Internet, as with previous innovations in transacting technologies, creates forces for agglomeration and disagglomeration. According to Storper (1997), agglomeration and disagglomeration forces are the two main opposing forces that shape the coordination over space that characterizes clusters and similar forms of localized economic specialization. Leamer and Storper (2001) state that the Internet can transform some coordination tasks into routine activities. Therefore some activities can take place over an extended geographical region, thus contributing to de-agglomeration. This is, for example, the case of software production outsourcing. On the other hand the Internet can create a number of new activities that can lead to agglomeration. However, Leamer and Storper (2001) believe that "coordination of new and innovative activities depends on the successful transfer of complex uncodifiable messages, requiring a kind of closeness between the sender and the receiver that the Internet does not allow" (Leamer and Storper, 2001). In a similar fashion, Maignan *et al.* (2003) conduct a theoretical and empirical analysis of the effect of ICTs on the centripetal and centrifugal forces that lead to agglomeration and disagglomeration of economic activities in spaces. They theoretically conclude that ICTs can contribute to both centripetal and centrifugal forces and that empirical studies are supporting both views. However, they empirically observe that the use of ICTs would mostly contribute to an increase in industrial geographical concentration and to a reduction of transport costs, thus supporting Leamer and Storper's (2001) arguments.

Other studies have showed the limited use of ICTs in clusters by, for example, pointing out the historical backwardness of industrial districts with regard to ICT adoption (for example Micelli and Di Maria, 2000). The seminal article by Kumar *et al.* (1998) shows how an inter-organizational system that had been adopted to support coordination among firms in the Prato textile cluster, located in Tuscany, Italy was abandoned ten years after its installation. Kumar *et al.* (1998) conclude that the inter-organizational system could not add any value to the existing forms of coordination of the cluster and that trust and personal relationships were the main coordination mechanisms. Similarly, Gottardi (2003) in an analysis of ICT implementation in industrial districts, which he characterized as economic and social homogeneous environments, concludes that ICTs are difficult to diffuse in industrial districts. This is mainly due to the presence of tacit knowledge and to the fact that ICT-based communication is not superior to pre-existing, interpersonal modes of communication.

Abecassis-Moedas and Caby-Guillet (2006) conducted a study to evaluate the influence of cluster sociological factors such as interpersonal trust on the implementation of ICTs in the cluster. To do this they specifically studied the adoption of Electronic Data Interchange (EDI) in the clothing industry in France and the USA. They found in general that EDI penetration in the clothing industry is quite low and it is higher in the US than France. EDI is mostly used in the cases of US-based large textile firms wanting to relocate their manufacturing to other regions such as Mexico by using IT. Abecassis-Moedas and Caby-Guillet (2006) find that EDI can be used for coordination when transactions are frequent and involve a high volume of goods. According to them, this allows standardization and thus the use of electronic networks such as EDI to support such transactions. Similarly to other studies (for example Kumar *et al.*, 1998) they also found that a pre-existing high level of trust among the exchange partners limits the implementation of ICTs because embeddedness and interpersonal communication are efficient pre-existing modes of communication.

Belussi (2005) in a study of the diffusion of Internet technologies in three important North East Italian districts found that firms "have preferably chosen relatively low cost 'ready-made' ICTs rather than more complex systemic technologies useful for electronic communication and interactive data exchange" (p. 263). In addition such technologies mostly supported transactions with end customers. Examples of the most used technologies are company-Internet web sites, e-mail, corporate banking, e-commerce, and collective Internet sites. The companies were reluctant to use business to business linkages with subcontractors and suppliers such as EDI and ERP. As Belussi (2005: 247) states "this should not be interpreted as a lock-in phenomenon, but as a sign that they rely on flexible and trustful informal communication that cannot easily and efficiently be virtualized in electronic form". Belussi (2005) concludes, however, that ICTs might have a bigger role in the future of industrial districts and clusters to support external linkages and distant interaction, for example, for organizing explorative R&D, absorptive technological spillovers, access to global pipelines and relocation of subcontracting chains.

Finally in a thorough literature review of ICT usage in industrial clusters, Steinfield and Scupola (2006) distinguish between two fundamental ways that a local ICT infrastructure might be used in business clusters. The first relates to connectivity within the cluster for coordination and collaboration purposes. This includes telework (for example Westfall, 2004), as well as ICT applications for information sharing and collaborative work within firms and across firms belonging to the same cluster. The second way in which an ICT infrastructure can be used in a business cluster is for electronic commerce transactions. These transactions can be either business to business or business to consumer. Steinfield and Scupola (2006) investigated the use of ICT infrastructure in a case of a well known Danish biotech cluster, Medicon Valley. Their results show that low cost broadband access had benefited both big and small firms for coordination and collaboration within the cluster. In addition firms in the cluster also practiced some form of e-commerce such as receiving inquiries through the web by distant customers or taking orders via e-mail. For SMEs this form of e-commerce was mainly Internet-based, thus confirming Belussi (2005)'s results that small and medium size companies prefer low cost, "ready-made" technologies.

As this review shows, the literature on ICTs in industrial clusters is still at the beginning and shows different, at times, contrasting roles for ICTs in the creation and evolution of clusters. Some authors state that ICTs are not important in clusters mainly due to tacit knowledge and interpersonal relation based on personal trust (for example Kumar *et al.*, 1998). However other recent studies do show that ICTs have a role in clusters and industrial districts. This is especially true for internal collaboration and coordination (Steinfield and Scupola, 2006) and for simple electronic commerce transactions such as managing relations with customers (Belussi, 2005; Steinfield and Scupola, 2006). Therefore from the above literature review we can conclude that ICTs can support the following activities in clusters: connectivity within and outside the cluster for coordination and collaboration purposes; electronic commerce transactions within and outside the cluster.

INNOVATION AND CLUSTER

The institutional environment created by governments in the form of policies and interventions is very important to foster innovation and therefore economic development of developed as well as developing nations. Often the economic performance of a country is influenced by the promotional and proactive policies of governments (for example Coase, 1990; North, 1990; Teubal, 1979). Sørensen and Levold (1992), for example, point out the importance of social and institutional factors such as the involvement of government and the role of the single entrepreneur as enablers and stimulators of innovation. Furthermore, several authors have found that the external environment, including industry associations and the government, is very important in the adoption and diffusion of technological innovations (for example Tornatzky and Fleischer, 1990; Teubal, 1979; North, 1990). Two major models linking innovation to local economic development are Porter's (1998) diamond model and Etzkowitz and Leydesdorff's (2000) Triple Helix model.

Porter's diamond model was developed to address the question of why a nation achieves international success in a particular industry. He identifies four major areas in which the state can contribute to cluster's development:

1. Create factor conditions including specialized education and training programs, support the establishment of local university research in cluster-related technologies, support cluster-specific information gathering and compilation, enhance specialized transportation, communications and other infrastructure.
2. Create a context for firm rivalry including elimination of barriers to local competition, organizing relevant government departments, a focus on efforts to attract foreign investment and export promotion around clusters.
3. Support related industries by encouraging cluster-specific efforts to attract suppliers and service providers from other locations, establish cluster oriented free-trade zones, industrial parks or supplier parks.
4. Create demand conditions including pro-innovation regulatory standards affecting the cluster to reduce regulatory uncertainty, and stimulating early adoption, such as by acting as sophisticated buyer of the cluster's products or services.

The Triple Helix model (Etzkowitz and Leydesdorff, 2000) instead focuses on three major actors (or helixes) as responsible for local development and innovation: the state, the industry and the academic world. The model stresses the importance of interaction among actors within the same helix and between different helixes to enhance innovation and local economic development. In addition, relationships between the three helixes should be interactive, embedded in projects and in shared values. In this model, the universities are envisioned as a driving force in economic development, besides research and education, together with the other actors.

WHY THE MEDICON VALLEY CLUSTER?

To illustrate how a network organization has used Internet-based technologies to increase cooperation and collaboration among actors located within a cluster and between the cluster and other regions of the world, we conducted a case study of the Medicon Valley cluster. Specifically we focused on the roles that the Medicon Valley Academy (MVA), through its use of ICTs, has had in the cluster's development and branding. This case was chosen first of all because Denmark has been the first country in the world to develop policy measures to create networks or clusters of companies similar to the practice-based networks considered to be uniquely linked to northern Italy (Rosenfeld, 2001). In addition as pointed out by Steinfield and Scupola (2006) this cluster was chosen for a number of other reasons. First, biotechnology is an increasingly important sector in many economies, and many attempts have been made worldwide to develop successful clusters in biotechnology (Cooke, 2002b). Consequently significant research has been conducted on biotechnology clusters (Audretsch, 2001; Audretsch & Stephan, 1996; Cooke, 2002a, 2002b; Frank, 2002; Wolff, 2003). Second, the Medicon Valley has been a highly successful example of a biotechnology cluster, achieving a prominent position especially at a European level (Frank, 2002; Wolff, 2003). Finally biotechnology is a global, knowledge intensive industry, placing more emphasis on information transfer than the transfer of physical goods (Cooke, 2002b; Powell *et al.*, 2002; Powell *et al.*, 1996). This offers great potential to reveal important uses of ICTs for coordination and cooperation within and outside the cluster.

BACKGROUND OF THE MEDICON VALLEY CLUSTER

The Medicon Valley cluster is one of the top European bioregions featuring cross-border partnerships between industry, universities, hospitals, science parks, investors and business service. The cluster has its historical roots in the dairy producers that for several centuries have characterized the region (Norus, 2004 in Lyck, 2006). The cluster is characterized by companies such as Carlsberg and Chr. Hansen, that in 1874 was the first company in the world to market an industrial enzyme (Lyck, 2006). Innovation is therefore a key competence in the cluster and takes place through cooperation among different actors such as hospitals, the pharmaceutical industry and universities (Erhvervsfremmestyrelsen, 2001). The companies located in Medicon Valley encompass all activities relating to, or dependent on biomedical research and development. The company types vary from very small, entrepreneurial start-up companies to large international companies such as Novo Nordisk, Lundbeck, Coloplast, Radiometer, BASF, and NeuroSearch (Boston Consulting Group, 2002). The cluster has seen a tremendous growth in recent years as the number of new biotechnology companies has increased significantly, with both new local companies as well as new subsidiaries of foreign biotechnology companies. Four main research-related core competences can be distinguished in the Medicon Valley cluster: diabetes, inflammation, neuroscience, and cancer (Lyck, 2006, www.mva.org). Another core competence of this region is within the service sector. The most important international competitors of this cluster are the world's leading biotech regions in Boston, San Francisco and San Diego in the USA, Cambridge in the United Kingdom, Bayern in Germany and Paris in France.

The Medicon Valley cluster is located in the Øresund Region. This region includes the Copenhagen area in Denmark and the southern region of Sweden called Scania. The Øresund region is becoming a very important regional center in Northern Europe. This is due to a variety of regional competences among which are good access to the Scandinavian and Baltic markets, a well developed infrastructure, a large international presence, high knowledge intensity, population (it is the largest metropolitan area in Scandinavia with a population of circa 3 million inhabitants), and a high concentration of companies (Table 10.1). The region has one of the highest

Table 10.1   Characteristics of Medicon Valley (adapted from the MVA home page, www.mva.org)

Population	Ca 3 million inhabitants – 22 percent of the total population of Denmark and Sweden
Number of universities	14
Number of university students	150 000
Number of scientists	5000 life science researchers
International accessibility	Copenhagen Airport has more than 15 million international passengers annually. There are direct flights to more than 125 cities in the world
Biotech/human life science activities in Medicon Valley	• 140 biotech companies with 3300 employees • around 70 pharma companies • around 130 medicotech companies • around 15 CROs (clinical research organizations) • 250 others (including service providers and investors)
Number of employees	35 000 employees in the biotech/pharma/ medicotech industry
Export	60 percent of all exports within life science from Sweden and Denmark originates from Medicon Valley
Number of hospitals	26 hospitals of which 11 are university hospitals
Science parks	7

per capita GDP worldwide with 29 100 euro per capita compared with 24 100 euro per capita on average in the European Union (OECD, 2002). Three important organizations are behind the promotion of the region: Region Skåne, Copenhagen Capacity and Medicon Valley Academy (www.mediconvalley.com). As already stated in this chapter, we focus on the MVA.

THE ROLE OF MEDICON VALLEY ACADEMY

Medicon Valley Academy (MVA) is a network organization within the biotech and life science area with the main purpose of supporting (that is taking care) of the cluster. Medicon Valley Academy was established at the end of 1996 and can be conceptualized as a concrete example of the triple helix model (Lyck, 2006). The main purpose of the organization is, in fact, to establish and increase collaboration among the different actors of the region, such as private companies, institutions, universities and support organizations, thus contributing to the development of the region (Etzkowitz and Leydesdorff, 2000). Medicon Valley Academy has also the purpose of developing the cluster and branding it internationally. This was clearly pointed out by a manager at Medicon Valley Academy:

> We are doing things for the development of the region and not only for our members . . . we would like the region to perform perfectly without us, but that is not the reality . . . we are trying to initiate things and we would like other organizations to do what we do, but we do not see any competitors. (Medicon Valley Academy Manager, June 2003)

Before the Medicon Valley brand was created, the Øresund region was just an agglomeration of companies, organizations and institutions specializing in biotechnology and related areas. According to the cluster policy that had been developed in Denmark in the 1990s (Rosenfeld, 2001; Erhvervsfremmestyrelsen, 2001; Wegloop, 1996) around 1996, policy makers and other experts started to discuss how to strengthen the Danish biotechnology sector (Lyck, 2006). Therefore the need arose to make it more visible. This was done by stressing the region's competences and by investing in research and development in the field of biotechnology, thus laying the basis for the formation of the Medicon Valley cluster. As a manager at Medicon Valley Academy stated:

> Initially university professors and other competent people tried to understand how to capitalize on the competences in the region and started thinking about how to give a brand and then they came up with the name of Medicon Valley,

there is Silicon Valley and so . . . (Medicon Valley Academy Manager, June 2003)

Medicon Valley Academy was created at the end of 1996/beginning of 1997 as a public project with the support of inter-regional funds from the European community (Lyck, 2006). In 2000, MVA was transformed into a member based organization with 50/50 public-private financing (through member fees) including all hospitals, universities and more than 200 companies in Medicon Valley. As an MVA manager stated:

> In the first three years we were a kind of public funded project from the European Union and Danish government and then in year 2000 we were transferred to a private member-based organization. (Medicon Valley Academy Manager, June 2003)

Today Medicon Valley Academy counts 13 employees that work "to improve the conditions for science and knowledge production, technology transfer, innovation and for the preconditions for enterprises to exploit this knowledge. In addition MVA works to visualize the potential of activities in both Sweden and Denmark as well as internationally" (www.mva.org). MVA is a not-for-profit organization and is managed by a Board of Directors and a staff located at offices in Ørestad City, Copenhagen and in the university town of Lund, Sweden. Medicon Valley Academy has an important role in supporting or "taking care" of the cluster. This is mainly done both by promoting the cluster internationally and by fostering innovation through knowledge creation, transfer and exchange both at local and international level. As an MVA manager said:

> It is my observation that the organizing capacity of the region is extremely important . . . So we positioned ourselves as the organizing capacity and we made a 3 point strategy mentioned in 3 words: to create, to transfer and to exploit knowledge. (Medicon Valley Academy Manager, June 2003)

How Medicon Valley Academy intends to support or "take care" of the cluster is clearly described by their mission statement:

> MVA will assist in optimising conditions for creating, developing and utilising knowledge in Medicon Valley thereby serving in the important role as facilitator for accomplishing the vision for Medicon Valley. By initiating, facilitating, and operating networking and knowledge-sharing platforms, as well as influencing regional and national decisions on the political or administrative levels.

The Medicon Valley Academy has the objective of contributing to the overall growth of the cluster and its establishment at international level.

MVA's vision is to have Medicon Valley among the five most attractive bioregions worldwide.

THE KEY ROLE OF INTERNET-BASED TECHNOLOGIES IN SUPPORTING THE CLUSTER

A number of cluster researchers have emphasized the importance of activities undertaken by government-sponsored associations in the promotion and the development of the cluster (Chiesa and Chiaroni, 2004; Cooke, 2002a; Turner, 2003). To achieve its mission to optimize "conditions for creating, developing and utilising knowledge in Medicon Valley thereby serving in the important role as facilitator for accomplishing the vision for Medicon Valley" the Medicon Valley Academy initiates, facilitates, and operates a number of networking and knowledge-sharing activities (Interview with MVA manager, June 2003; www.mva.org). These activities aim at connecting the actors within the region, branding the cluster internationally and establishing collaboration within the region, in Scandinavia and worldwide. Examples are the organization of annual biotech meetings, seminars, conferences, the publication of scientific reports and a monthly newsletter that is distributed to the members usually by e-mail.

Internet-based technologies play an important role in the way Medicon Valley Academy supports or "takes care" of the region. They are an integral part of MVA's strategy. This is clearly shown by the following statement:

> The agenda for us (is) to make an attractive region . . . and the means are network activities such as meetings, conferences, seminars plus the Internet facilities that we provide and other things such as a call centre so that people can call and ask where can we find such and such competences . . . The other thing we want to do is to create good conditions to exploit and commercialize knowledge. You know knowledge is out there, in many places, but we put it together . . . We achieve economy of scale because we collect the knowledge and there are a lot out there that use it . . . We do it cheaper sometimes. (Medicon Valley Academy Manager, June 2003)

Medicon Valley Academy uses especially Internet-based technologies and the World Wide Web (WWW) for coordination and collaboration within and outside the cluster. They also use such technologies for cluster branding, which is mainly done through their web page. For example, for coordination inside the cluster (for example Steinfield and Scupola, 2006) they use e-mail and search engines to gather information about the region and use the web page to make it available to potential users. As a Medicon Valley manager stated:

Internet helps a lot in our activities because we are collecting information for a lot of actors in the region and we put them in one place, we build a square for the region, a kind of interactive square, where you can find information about courses, new publications, and it is up to the network what kind of info they need and then they can go in and find it, . . . the Internet is perfect for providing info. (Medicon Valley Manager, 2003)

We use the web page to present our publications, annual report of the activities of the region, we use the Internet to collect information, for example when we had to put together our financing guide we contacted the people and asked them to give us the information or to check the information we had, if that worked that was ok, otherwise we had to call by phone. (Medicon Valley Manager, 2003)

MVA also acts as a facilitator for companies searching for a partner in the region or wanting to establish themselves in the region and their web site is very important in this, as showed by the following statement:

We are (referring to the web site) a global point of entry to the region. (Medicon Valley Manager, September 2006)

Despite most of the literature on ICTs in clusters (for example Kumar *et al.*, 1998) Internet-based technologies contribute also to build and maintain personal networks, as the following statement shows:

Our web page is very important to offer information about the sector, how to find financing, . . . how to do networking. (Medicon Valley Manager, 2003)

Internet and the World Wide Web are also extremely important in supporting another key objective related to Medicon Valley Academy's mission and the goal of cluster innovation: knowledge transfer. The literature on knowledge management (for example Hansen *et al.*, 1999) shows that explicit knowledge that can be stored, for example, in files or reports, and can easily be transferred using ICTs. Face-to-face meetings and other types of proximity-dependent arrangements are often considered necessary to facilitate the exchange of tacit knowledge (Nonaka, 1994). Similarly an MVA's manager states that ICTs are extensively used in Medicon Valley for the transfer of explicit knowledge both locally and globally, as showed by the following statement:

They (the companies) use a lot of IT for knowledge transfer and communication and these alliances across continents are increasing more and more. People meet and then they can continue the dialogue on the Internet, other times they have to meet each other again and so . . . But in between the Internet is extremely important for that knowledge transfer. (Medicon Valley Manager, 2003)

To collect and store the knowledge available in the region, MVA is also in the process of developing a web portal giving access to a database mapping the competences of the region. These competences include projects in the pipeline, company information and so on. They also intend to explore the opportunity of virtual networks (Interview with MVA Manager, September 2006). Moreover, just as innovative firms can use their connections to a global ICT infrastructure like the Internet to bring new knowledge into a region, which may then diffuse among local trading partners (Hicks and Nivin, 2000; Zaheer and Manrakhan, 2001), the MVA is also using the Internet to collaborate with clusters located all over the world. Examples are Scanbalt in Eastern Europe and the UK-Medicon Valley Challenge Program. This latter program is an initiative to increase research collaboration between the UK and Medicon Valley biotech research communities, in order to enhance overall competitiveness. Internet-based technologies have an important role in these collaborations among clusters/regions, as an MVA manager states:

> Biotech clusters are global and local at the same time . . . So ICTs are needed all the time both for local and global collaboration. (MVA Manager, September 2006)

CONCLUSIONS

This chapter has showed how Internet-based technologies can contribute to and support the development of localized economies such as industrial clusters, while contributing to the globalization of the economy by connecting companies and clusters of companies across different regions of the world. To do this the chapter has investigated the Medicon Valley cluster. Medicon Valley is a biotechnology cluster, located in Denmark and Southern Sweden, that is becoming one of the most important biotech clusters in Europe and worldwide (Frank, 2002; Wolff, 2003). In particular, the analysis has focused on how a network organization with a central position in the cluster, Medicon Valley Academy (www.mva.org), has used Internet-based technologies to increase collaboration and knowledge exchange among actors located inside the cluster and with actors located outside the cluster. The chapter also shows how Internet-based technologies have been used to increase cluster visibility and branding. In addition, by referring to the theme of this book which is "Innovation with care," the chapter has revealed how Medicon Valley Academy supports or "takes care of" the cluster by undertaking a number of activities to establish the cluster brand, promote the cluster internationally and attract international labor and companies to the region. The analysis has mainly been based on two interviews with employees of Medicon Valley Academy (MVA), its web page and other documentary material. We can conclude from this analysis that Internet-based technologies can successfully be used to support cooperation and collaboration within the same cluster and between clusters located in different regions of the world. However, we have to be careful about generalizing these results to any cluster. First, these conclusions are based on the study of only one cluster. Second the cluster is in biotechnology, which is a global, knowledge intensive industry, placing more emphasis on information transfer than the transfer of physical goods (Cooke, 2002b; Powell *et al.*, 2002; Powell *et al.*, 1996). Therefore these types of clusters offer great potential to reveal important uses of ICTs for coordination and cooperation within and outside the cluster (Steinfield and Scupola, 2006). Lastly, the cluster had been the result of a Danish policy initiative to strengthen the biotech sector in Denmark (Erhvervsfremmestyrelsen, 2001; Lyck, 2006).

REFERENCES

Abecassis-Moedas, C. and L. Caby-Guillet (2006), "ICT and trust in clothing clusters," paper presented to a workshop on the role of ICTs in interfirm networks and regional clusters at the International Conference on the Design of Cooperative Systems, Marseille, France, 9-12 May.

Asheim, B., P. Cooke, and R. Martin (2006), "The rise of the cluster concept in regional analysis and policy: a critical assessment," in B. Asheim, P. Cooke and R. Martin (eds), *Clusters and Regional Development – Critical Reflections and Explorations*, London and New York: Routledge.

Audretsch, D. (2001), "The role of small firms in US biotechnology clusters," *Small Business Economics*, **17** (1,2), 3–15.

Audretsch, D. and P. Stephan (1996), "Company scientist locational links: the case of biotechnology," *American Economic Review*, **86** (3), 641–52.

Becattini, G. (1990), "The Marshallian industrial district as a socio-economic notion," in F. Pyke, G. Becattini and W. Sengenberger (eds), *Industrial Districts and Inter-firm Cooperation in Italy*, Geneva: International Institute for Labor Studies, pp. 37–51.

Belussi, F. (2005), "Are industrial districts formed by networks without technologies? The diffusion of Internet applications in three industrial clusters," *European Urban and Regional Studies*, **12** (3), 247–68.

Boston Consulting Group (2002), "Commercial attractiveness of biomedical R&D in Medicon Valley," report prepared for Copenhagen Capacity and Position Skåne, accessed at www.mva.com.

Cairncross, F. (1997), *The Death of Distance*, Boston, MA: Harvard Business School Press.

Chiesa, V. and D. Chiaroni (2004), *Industrial Clusters in Biotechnology: Driving Forces, Development Processes and Management Practices*, London: Imperial College Press.

Cooke, P. (1998), "Introduction: origins of the concept," in H. Braczyk, P. Cooke and M. Heidenreich (eds), *Regional Innovation Systems*, London: UCL Press.

Cooke, P. (2001), "Regional innovation systems, clusters, and the knowledge economy," *Industrial and Corporate Change*, **10** (4), 945–74.

Cooke, P. (2002a), "New media and new economy cluster dynamics," in L. Lievrouw and S. Livingstone (eds), *The Handbook of New Media*, London: Sage, pp. 287–303.

Cooke, P. (2002b), "Regional innovation systems: general findings and some new evidence from biotechnology clusters," *Journal of Technology Transfer*, **27** (1), 133–45.

Coombs, R., K. Green, A. Richards and V. Walsh (1998), *Technological Change and Organization*, Cheltenham, UK and Lyme, USA: Edward Elgar.

Crouch, C., P. Le Galès, C. Trigilia and H. Voelzkow (2001), *Local Production Systems in Europe: Rise or Demise?*, Oxford: Oxford University Press.

Dosi, G. (1988), "The nature of the innovative process," in G. Dosi, Chr. Freeman, R. Nelson, G. Silverberg and L. Soete (eds), *Technical Change and Economic Theory*, New York: Pinter Publishers, pp. 221–38.

Etzkowitz, H. and L. Leydesdorff (2000), "The dynamics of innovation: from national systems and 'mode 2' to a triple helix of university-industry-government relations," *Research Policy*, **29** (2), 109–23.

Erhvervsfremmestyrelsen (2001), *Kompetenceklynger i dansk erhvervsliv-en ny brik i erhvervspolitikken*, Copenhagen: Erhvervsfremmestyrelsen.

Frank, L. (2002), "Biotechnology in the Medicon Valley," in *Nature Biotechnology*, **20**, 433–6.

Gottardi, G. (2003), "Why do the ICT and the Internet find it hard to spread into industrial districts and favour knowledge exchanges?" in F. Belussi, G. Gottardi and E. Rullani (eds), *The Technological Evolution of Industrial Districts*, Boston, MA: Kluwer.

Hansen, M.T., N. Nohria and T. Kierney (1999), "What's your strategy for managing knowledge?", *Harvard Business Review*, **77** (2), 106–16.

Hicks, D.A. and S.R. Nivin (2000), "Beyond globalization: localized returns to infrastructure investments," in *Regional Studies*, **34** (2), 115–27.

Krugman, P. (1998), "What's new about the new economy geography," *Oxford Review of Economic Policy*, **14** (2), 7–17.

Kumar, K., H van Dissel and P. Bielli (1998), "The merchant of prato – revisited: toward a third rationality of information systems," *MIS Quarterly*, **22** (2), 199–226.

Leamer, E.E. and M. Storper (2001), "The economic geography of the internet age," *Journal of International Business Studies*, **32** (4), 641–65.

Loasby, B.J. (1998), "The concept of capabilities," in N.J. Foss and B.J. Loasby (eds), *Knowledge, Institutions and Evolution in Economics*, London, Routledge, pp. 163–82.

Lundvall, B.-A. (1993), "Explaining interfirm cooperation and innovation: limits of the transaction-cost approach," in G. Grabher (ed.), *The Embedded Firm – on the Socio- economics of Industrial Networks*, London: Routledge, pp. 52–64.

Lyck, L. (2006), *Øresundsregionalisering – Medicon Valley*, Frederiksberg: Copenhagen Business School, Centre for Tourism and Culture Management.

Maignan, C., D. Pinelli and G. Ottaviano (2003), "ICT, cluster and regional cohesion: a summary of theoretical and empirical research," *Nota Di Lavora* (58), accessed at http://papers.ssrn.com.

Malmberg, A. and P. Maskell (2002), "The elusive concept of localization economies – towards a knowledge-based theory of spatial clustering," *Environment and Planning*, **34** (3).

Malmberg, A. and D. Power (2006), "True clusters: a severe case of conceptual headache," in B. Asheim, P. Cooke and R. Martin (eds), *Clusters and Regional Development – Critical Reflections and Explorations*, London and New York: Routledge.

Marshall, A. (1936), *Principles of Economics: An Introductory Volume*, London: McMillan.

Martin, R. and P. Sunley (1998), "Slow convergence? The new endogenous growth theory and regional development," *Economic Geography*, **74** (3), 201–27.

Maskell, P. and L. Kebir (2006), "What qualifies as a cluster theory?" in B. Asheim, P. Cooke and R. Martin (eds), *Clusters and Regional Development – Critical Reflections and Explorations*, London and New York: Routledge.

Matthiessen, C.W. and A.W. Schwarz (1999), "Scientific centres in Europe: an analysis of research strength and patterns of specialisation based on bibliometric indicators," *Urban Studies*, **36** (3), 453–77.

Micelli, S. and E. Di Maria (eds) (2000), *Distretti industriali e tecnologie di rete. Progettare la convergenza [Industrial districts and network technologies]*, Milan: Franco Angeli Ed.

Nonaka, I. (1994), "A dynamic theory of organizational knowledge creation," *Organization Science*, **5** (1), 14–37.

North, D.C. (1990), *Institutional Structure and Institutional Change*, New York: Cambridge University Press.

Organisation for Economic Co-operation and Development (OECD) (2002), *OECD Science, Technology and Industry Outlook 2002*, Paris: OECD.

Piore, M. and C. Sabel (1984), *The Second Industrial Divide*, New York: Basic Books.

Porter, M.E. (1990), *The Competitive Advantage of Nations*, New York: Free Press.

Porter, M.E. (1998), "The Adam Smith address: location, clusters, and the 'new' microeconomics of competition," *Business Economics*, **33** (1), 7–13.

Powell, W.W., K.W. Koput, J.I. Bowie and L. Smith-Doerr (2002), "The spatial clustering of science and capital: accounting for biotech firm-venture capital relationships," *Regional Studies*, **36** (3), 291–305.

Powell, W.W., K.W. Koput and L. Smith-Doerr (1996), "Interorganizational collaboration and the locus of innovation: networks of learning in biotechnology," *Administrative Science Quarterly*, **41** (1), 116–45.

Rogers, E.M. (1995), *The Diffusion of Innovations*, 4th edn, New York: Free Press.

Rosenfeld, S.A. (2001), "Networks and clusters: the yin and yang of rural development," *Proceedings Rural Conferences, 2001* (September), 103–20.

Scott, A.J. (1988), "Flexible production systems and regional development. The rise of new industrial space in North America and Western Europe," *International Journal of Urban and Regional Research*, (12), 171–86.

Sørensen, K.H. and N. Levold (1992), "Tacit networks, heterogenous engineers, and embodied technology, science," *Technology and Human Values*, **17** (1), 13–35.

Steinfield, C. and A. Scupola (2006), "Explaining ICTs infrastructure and E-commerce uses and benefits in industrial clusters, evidence from a biotech cluster," in R. Sprague, Jr. (ed.), *Proceedings of the 39th Hawaii International Conference on System Sciences*, IEEE Computer Society.

Storper, M. (1997), "Regional economies as relational assets," in R. Lee and J. Willis (eds), *Society, Place, Economy: States of the Art in Economic Geography*, London: Edward Arnold.

Teubal, M. (1979), "On user needs and need determination, aspects of the theory of technological innovation," in M. Baker (ed.), *Industrial Innovation, Technology, Policy, Diffusion*, London: Macmillan, pp. 266–93.

Tornatzky, L.G. and M. Fleischer (1990), *The Processes of Technological Innovation*, Lexington, MA: Lexington Books.

Turner, C. (2003), "New media industries: a quantitative research study," paper presented at a workshop on business clusters at the International Conference on Communities and Technologies, Amsterdam, The Netherlands, 19-21 September.

Wegloop, P. (1996), "Innovation firm strategy and government policy: why and how they should be linked", Roskilde University Department of Social Sciences, PhD thesis no. 10/1996 ISSN No. 0909-9174, March.

Westfall, R. (2004), "Does telecommuting really increase productivity," *Communications of the ACM*, **47** (8), 93–6.

Wolff, M.F. (2003), "Biomedical cluster blossoms in Scandinavia", *Research Technology Management*, **46** (4), 2–4.

Zaheer, S. and A. Manrakhan (2001), "Concentration and dispersion in global industries: remote electronic access and the location of economic activities," *Journal of International Business Studies*, **32** (4), 667–86.

11. The "Mad Max Puzzle": positioning and the lone inventor

Jerome Davis and Lee N. Davis

Individuals, much like companies and other institutions, must create relationships across actors with differing starting points and approaches to innovation. This chapter examines the problem of the lone inventor desirous of getting his invention to market without its value being unreasonably appropriated by a "partner" (here Big Widget Inc.), which we term the "Mad Max Puzzle." It is generally assumed that independent inventors, whether their inventions are patented or unpatented, will be cheated by the Big Widgets of this world. The argument is that since Big Widget has the capability to appropriate this value, she will necessarily do so. We disagree. We begin our analysis with "Mad Max," the inventor of the proverbial mousetrap. We demonstrate the importance of positioning with respect to time, with respect to third (or fourth) parties, and with respect to the negotiation agenda. In conclusion, we examine the well publicised case of Robert Kearns, a lone inventor whose invention of the intermittent windshield wiper was appropriated by the Ford Motor Co., to pinpoint how through positioning himself better, he might have avoided a lifetime of lawsuits.

INTRODUCTION

Every capitalist society has its mythical lone genius inventor (we will call him "Mad Max"). Mad Max invents an ingenious mousetrap, machine, whatever (we will call it a "widget"). Mad Max lacks the resources to develop his widget, or sometimes even to get it patented (if it is patentable). He approaches a firm to pay him for his idea or set him up in business. Mad Max either intentionally or accidentally reveals his idea. The firm appropriates the idea without paying for it, develops, and markets the widget, pocketing the associated rents. Max dies impoverished, claiming he "wuz robbed." This puzzle, the "Mad Max puzzle," combines various well known economic problems: the economic properties of information, the need to secure appropriability, and the problems of a co-ordination game with various Nash equilibrium strategies.

Several scholars have explored the situation of the independent inventor. Udell (1990), and Parker and Udell (1996), for example, discuss the special characteristics of these inventors. Three important empirical studies have illuminated aspects of the commercial success of independent inventors. Åstebro (1998) found that inventions by independent inventors survive for about as long as the average start-up, and gross profit margins are comparable to the pharmaceutical industry (29 percent). But the probability of inventions by independent inventors that reach the market is 6.5 percent. This percentage is four to eight times less than for inventions developed by established firms. Dahlin *et al.* (2004), in a study of 225 tennis racket patents granted in the US between 1981 and 1991, determined that independent inventors are a heterogeneous group who generate inventions that are overrepresented both among the most impactful and the least inpactful patents. Åstebro and Dahlin (2005) analysed the role of technological opportunity on independent inventors' decisions to patent and commercialize inventions. Their most important finding was that technical feasibility (the likelihood that an invention is technically sound and complete) appears important in affecting patenting but unimportant in affecting commercialization. In other words, inventors patent inventions without much thought of commercialization conditions.

Our chapter differs from this work in that we attempt to explain the reasons behind the frequent failure of many independent inventors to successfully realize the commercial value of their inventions commercially, seen in the context of positioning strategies. Working off (a) the literature on patenting and appropriability (for example, Mazzoleni and Nelson, 1998; L. Davis, 2004) and (b) the concept of strategic positioning (see for example Schelling, 1960; Bazerman and Neale, 1992; Camerer, 1997), as developed in this volume,[1] we examine ways of mitigating the Mad Max problem. First, we consider the Mad Max puzzle in simple game theoretic terms. Then, we elaborate our positioning approach, one of the "innovation with care" themes in this book, and apply this to see how one might solve the problem faced by both Mad Max and the potential buyer of his idea(s), whom we call Big Widget Inc. Finally, we apply our framework to a real world case and famous example of the Mad Max puzzle, Robert Kearns' battle over the rights to his invention of the "intermittent windshield wiper." His invention was appropriated first by the Ford Motor Company and later by the entire automobile industry. It is our contention that had Kearns followed the positioning strategy outlined here, his odds of receiving licensing fees and avoiding a lifetime of costly lawsuits against American and international automobile manufacturers would have been considerably improved.

MAD MAX AND BIG WIDGET: THE FUNDAMENTAL DILEMMA

In initially setting up our Mad Max puzzle, we make several assumptions. The first is that Mad Max has been granted a patent on the invention, giving him the exclusive rights to make, sell or use the invention for 20 years. This is because possession of a patent will *ceteris paribus* put Max in a stronger position than he would be without one (the patent provides the guarantee that no one else has a patent on that invention, and gives legal protection against infringement). Nevertheless, the patent law also requires that the patent document is published after 18 months from the date of the application, making that information freely available to others to read, learn from, and improve upon. The second assumption is that Mad Max possesses certain secret know-how that is critical to making the invention work, but is not included in the patent application. While, an inventor, to receive a patent, must demonstrate that the invention is novel, non-obvious, and industrially useful, and described it so completely and precisely that an expert "skilled in the art" can implement it, inventors routinely leave out key bits of know-how and keep it secret. While this know-how isn't necessary to achieve patent protection, without it, the invention will not function very well in practice. Without access to this secret know-how, Big Widget cannot commercialize the invention successfully. Third, we assume that neither party has any substantive information about the other prior to their negotiation, and as a result, no expectations as to what their behaviour might be. (We will relax some of these assumptions in later sections.)

There are three possible outcomes to our puzzle:

1. In his negotiations with Big Widget, Mad Max withholds the critical secret know-how required to make the invention work and makes no sale, but gains more general information (for example, information about other inventions that exist; other companies that are engaged in similar research, and the like) that he may use in later negotiations, either with Big Widget or alternative interested parties.
2. Mad Max accidentally or intentionally reveals the critical know-how (first move), and Big Widget uses this to develop the invention commercially herself, pocketing the rents without compensating Mad Max.
3. Mad Max licenses out patent rights to the invention to Big Widget, to develop it commercially. In the first move, Big Widget gives a binding pledge that she will not steal the idea solely for her own benefit. As a result, both mutually profit from the invention.

Mad Max

	Reveal know-how	Not reveal know-how
Not steal	5 5	1.5 1
Steal	0.5 9.5	1.5 1

Big Widget (row label)

Figure 11.1   Initial positioning: Mad Max and Big Widget Inc.

Figure 11.1 illustrates the positioning behind these strategies. To the left we have the position of Big Widget going into the negotiations. Big Widget either intends to steal or not steal the invention Mad Max is selling, but must have access to the critical secret know-how. Across the top are Mad Max's possible positions. He will either reveal his critical know-how to Big Widget or not reveal it. Note that this figure illustrates the problem in the absence of Mad Max's calculations as to the probability that Big Widget will cheat and steal his invention, or Big Widget's calculations that Mad Max's invention is impractical and cannot be commercialized. Each of the boxes represents a set of payoffs. Those values to the upper right of the boxes represent Mad Max's payoffs, those to the lower left Big Widget's payoffs.

What is interesting, here, is that Mad Max always gains through this exercise. If he does not reveal sufficient know-how behind his idea (so that irrespective of Big Widget's intention to appropriate the invention, it is impossible for her to do so), Mad Max benefits from the learning experience 0.5, and by the retention of the option to reveal his know-how to another prospective buyer or to Big Widget in another context (1 + 0.5 = 1.5 in the upper right hand corners of the two right hand boxes). He even receives a payoff in the worst possible case for him, where he reveals his know-how and Big Widget steals the invention, 0.5 for the learning experience. (Note that we assume similar "learning payoffs" of 1 for Big Widget.)[2] Of interest are two particular variants of this game.

First, assume that Big Widget and Mad Max move simultaneously. Under these circumstances, Mad Max will assume that Big Widget expects to receive 9.5 or 1 if she tries to steal the invention, and only 5 or 1 if she does not intend to do so. Big Widget will realize this and assume that Mad

Max will get either 5 or 0.5 if he reveals his secret know-how, and only 1.5 if he does not. Given that Mad Max stands to gain 1.5 with certainty if he does not reveal the know-how, and that he calculates that Big Widget prefers a return of 9.5 to 5, it will always pay Mad Max not to reveal his secret, but to engage Big Widget in negotiations. Given that Big Widget makes a similar calculation, and that she will still gain 1 if she negotiates with Mad Max, it will pay her to engage Mad Max in negotiations even if Mad Max does not reveal his secret.[3] We have standoff, where it pays both parties to negotiate, but not to enter into any commercial agreement. Arguably, in fact, the more Big Widget has invested in related R&D in the area of Max's widget, the less likely she will be to cooperate. (This is because extensive R&D reduces the odds that an outsider has anything constructive to add to what is already known.)

The problem is that it is as unwise for Big Widget to precommit before she is privy to the secret as it is for Mad Max. The reason is less that Big Widget will intentionally "cheat" Mad Max (although this remains a very real possibility, as we shall see) but more that the nature of her uncertainty is different. Let us assume that Mad Max expects that there is a 50-50 chance that Big Widget may appropriate his invention, he can incorporate this into his strategy. This means that Big Widget must not only take into account the possibility that Mad Max may "fudge" his information, but also the possibility that Max's invention has no commercial value – irrespective of whether or not she decides to steal it.

Let us assume that the Big Widget assesses that the probability that Max's invention will be a commercial success is 0.1 and that Mad Max assesses the probability of his obtaining his desired result 0.5. These altered payoffs are illustrated in Figure 11.2. Note that the values of the learning experience are not affected by these probabilities.[4] Mad Max would expect a payoff of 2.75 ((0.5*4.5) + 0.5) should the two parties cooperate. Failure to reveal his know how would yield 1 ((1*0.5) + 0.5 = 1), and should he reveal and Big Widget steal, he would get 0.5. Big Widget would receive 1.85 ((8.5*0.1) + 1) should she steal Max's revealed know-how, 1.5 should she cooperate, and 1 for all other variations. The problem is one of differing reduced pay-off expectations. Big Widget's expectations are significantly lower than are Mad Max's. Looking at it from Big Widget's point of view longer does stealing Max's invention mean a 9.5 gain; it is now 1.85. Note however that Mad Max does not make the same calculation, as he calculates his odds are 50–50. (This reduces his pay-offs to 3.25, 1, and 0.5 respectively). Similarly, Max's argument that Big Widget should precommit to appropriate in such a way that he is guaranteed a cash payment of 2.25 for his know-how could be seen by Big Widget as being unfair.[5]

Mad Max (*p* = 0.5)

	Reveal know-how	Not reveal know-how
Not steal	2.75 1.5	1 1
Steal	0.5 1.85	1 1

Big Widget (*p* = 0.1)

*Figure 11.2   Mad Max and Big Widget Inc. The impact of expectations:
risk-weighted payoffs (*p* = probability of success)*

An additional problem lies in the fact that a precommitment agreement not to appropriate can be a legally binding on Big Widget, while it is difficult to bind Max legally to the commercial validity of his idea, where the idea is imperfectly known to Big Widget, and Max is reluctant to disclose any further information for fear that Big Widget will appropriate it.

Big Widget has a further option: to try to "invent around" Mad Max's patent and take out its own patent/s on a related invention. The greater the degree that Big Widget has an ongoing R&D program in a similar area, and determines that there is room enough for both Mad Max's idea and its own, the greater the benefits of such inventing around. Thus if the signs are increasing that Max is tempted to fudge, it will prompt the Big Widget response, "Don't call us. We will call you."

What is usually depicted as a simple game of Big Widget's exploiting the unworldly but inventive genius has in fact become a complicated coordination game, where both sides are positioning themselves before either is willing to commit, and where it doesn't pay the buyer to commit pending the availability of information which the seller will not disclose for fear of its being stolen by the buyer.

POSITIONING OVER TIME

The timing of Mad Max's approach and release of information plays a major role in positioning here. Assume that Big Widget sees Mad Max either as having a commercially valuable invention (*p* = 0.15),[6] or a worthless idea (*p* = 0.85) and negotiations are proceeding on the basis of payoffs based on the assumptions behind Figure 11.1. Big Widget cannot tell in advance whether Mad Max has a valuable idea or not, partly because she does not know about Max's secret know-how, partly because no one can really know how well an invention will do once commercialized, and partly – the focus of this section, because she does not know what type of agent Max is. Based on her initial impressions, she may agree to meet Max. This will depend on how the players sequence their moves. In any case, Big Widget has the following considerations regarding a possible meeting with Mad Max.

First, since the probability that Mad Max has a commercially valuable invention is only 0.15, he is making a mistake in approaching Big Widget. Therefore, Big Widget should reject his approach, since his idea is not worthwhile.

Second, while Big Widget realizes that even though the chances of success are just 15 percent, based on her previous experience with other inventors, she intuitively believes that Max's idea is so good that it will beat the odds. Under this intuitive criterion, Big Widget will meet him and listen to his proposal.

Third, the better Mad Max's idea is, the lower the odds that he will reveal the critical know-how not included in the patent. This worsens the 15 to 85 odds that Big Widget will find negotiation profitable. She should not meet Max and listen to his proposal.

It should be noted in this context that instinctively, Big Widget should not be willing to meet Mad Max at all, since in two of the three above rationales, Mad Max has nothing to offer. (This might in fact explain some of the reluctance that many real life firms have in meeting lone inventors with "valuable" ideas.)

It is in this context that Mad Max has to position himself. First, he must convince Big Widget that while Big Widget feels that the odds of his invention being commercial are only 15 percent, his idea is so good it can beat the odds. Second, he must position himself in such a manner that Big Widget feels confident in his *bona fides*, and will not resort to stealing his idea.

Timing can help Mad Max partially achieve both objectives. We will illustrate our problem in terms of two time lines. Time Line One (Figure 11.3) represents the situation as described to this point. The only condition under which the inventor will find it in his interests to commit resources, in this case reveal his secret know-how, is that Big Widget herself commits resources to the invention prior to his revealing the know-how.[7] In our game thus far, this has not been the case. Big Widget can obtain her best solution appropriating the invention. Her second best solution would be cooperating with Mad Max. For Mad Max, however, cooperation is the

Inventor commits resources
(reveals secret know-how)

t

t_1 t_2

Potential purchaser
commits resources

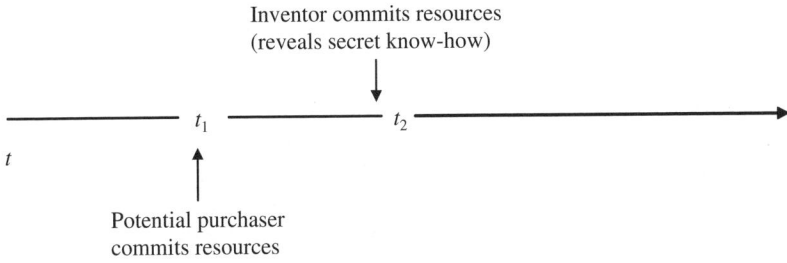

Figure 11.3   *Time line one: two step sequencing*

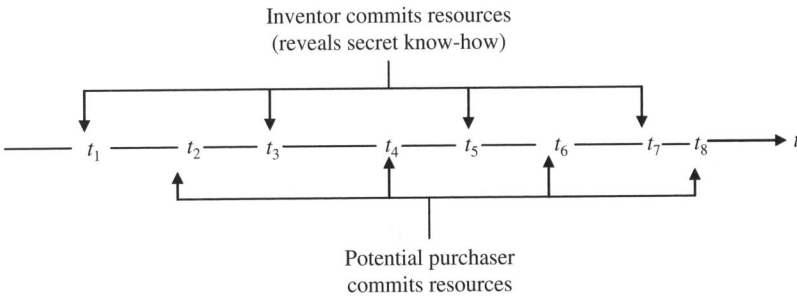

Inventor commits resources
(reveals secret know-how)

t_1 t_2 t_3 t_4 t_5 t_6 t_7 t_8 t

Potential purchaser
commits resources

Figure 11.4   *Time line two: multiple step sequencing*

preferred option as his assessment of success is 50 percent. So Big Widget loses nothing by waiting Mad Max out.

Time Line Two (Figure 11.4) represents another positioning solution to the problem. Here, the inventor divides his know-how up in a piecemeal fashion, the different pieces to be revealed at t_1, t_3, t_5, and t_7. These moves are countered by the purchaser, who gradually commits resources in exchange for the pieces of the idea. In this manner, both players can commit and receive resources in turn. (Clearly, there are variations on this time line. For example, both inventor and interested firm could commit resources simultaneously, and then check on the value of what they have received before going on to the next step.) The problem with this solution is that it is not amenable to all forms of know-how. Some know-how cannot be released piecemeal. Others can, but the "shape and size" of the missing individual pieces can be guessed at by the buyer so that there is little point in continuing the exercise.[8]

While the problems illustrated by Time Lines One and Two can be real enough, there are other facets of the puzzle which minimize these problems. Positioning can lead to commitment of resources so that Mad Max's payoff can be made in the future, so that the "decomposition" of the problem in Time Line Two becomes less necessary. For example, the two parties can send various subtle signals to each other. Big Widget might pay for a pre-liminary development plan. The two parties might sign a licensing contract. This assures Mad Max of a return for his idea in the future (should it be commercialized), and reduces Big Widget's immediate out-of-pocket expenses. Big Widget could also make an upfront payment to Mad Max for the right to continue to explore the possibilities of cooperation, or pur-chase a license option.[9] Similarly, if our conclusions about reciprocity are correct, the sequencing of idea bits and Big Widget's piecemeal commit-ment become less important. But in each of these instances, Mad Max's positioning strategy is vital to his success.

POSITIONING WITH A MULTITUDE OF AGENTS

A common misconception of the lone inventor is that he is alone. In fact, independent inventors have access to multitude of agents, with different expertise. Max must confront a multiplicity of tasks.

- apply for a patent;
- ascertain whether any competing patent (or patents) exists that render his patent invalid (if there are none, he must be sure that his patent is written in such a manner that his claims are incontestable, if there are, he must determine how to differentiate his invention enough from the existing patent/s so as to obtain a new patent);
- develop a prototype of his invention or refine a prototype to certain specifications to satisfy potential buyers;
- locate potential buyers, and acquire as much information about them and their reputations as he can;
- learn how to negotiate with potentially interested parties;
- take out insurance against possible future patent infringement; and
- investigate licensing choices (for example, should he try for an exclu-sive license agreement with one firm, or several non-exclusive ones?).

Finally, he must provide finance and time for all of these activities. In practice, it will be close to impossible for the lone inventor to accomplish all these tasks himself in such a way that he can position himself advanta-geously in negotiations with any potential Big Widget.

Fortunately, there are a multitude of agents who, in pursuing their own interests, can aid Mad Max in his positioning strategy. First, Big Widget may not be the only interested party. Generally speaking, it might be better to approach a smaller widget-making firm which has not put as many resources into widget R&D, and may be more open to the radical invention Mad Max is trying to sell. Irrespective of the Big Widget's estimation of the odds in Figure 11.2, her interest in negotiation is in part predicated on his getting the critical know-how from Mad Max (positively valued at 1). This figure could well be lower than 1. It could even be -1 or lower! (Remember the bigger the firm, the better its engineering and R&D staffs, the more Big Widget will tend to see her own know-how as sufficient, and thus that negotiation with Mad Max as a net cost rather than a gain, which will further reduce her incentive to negotiate in good faith.) Other firms, whether or not they produce widgets, might be interested in diversifying into the area of Mad Max's invention. There is a big difference in negotiating with Big Widget, which has sunk capital in the manufacture of widgets and specific ideas as to how widgets should be designed, and a venture capital firm which can be relied upon to set Mad Max in business himself. By varying his approaches and discussing his invention with various firms, Mad Max is not only acquiring valuable knowledge by positioning himself more advantageously with the individual potential buyer. Such competition both strengthens Mad Max's position and enables him to assess the value of his invention as seen by various potential licensees.

Second, there are a host of invention service or invention "promotion" firms which, for a fee, can offer everything from marketing plans to courses on license negotiations, how to display a prototype, how to search patent databases, how to find patent agents, how to find licensing agents, and how to insure one's patented invention from infringement. Alternatively, an invention service firm can actively promote Mad Max's invention for him (for a fee, of course).[10] Patent agencies can perform patent searches, map out patentable areas, and apply on Mad Max's behalf. Applying for a patent is so complex that Mad Max would be wise not try to do this work himself. The patenting process not only strengthens Max's position *vis-à-vis* potential licensees, but also indirectly informs him of some of the less obvious pitfalls he might encounter. Bank loans often required for finance could also involve discussions with bank officers familiar with the pitfalls faced by the sole inventor. If they should extend Max a loan, it is in their interests that his positioning is as strong as possible.

Finally, there are likely to be government organizations aimed at regional development, or the development of technological niches or specialties, that are seen as desirable from a political point of view. They too can provide Max with positioning strength.

Such processes not only strengthen Max's position, but also work to convince potential licensees of Max's *bona fides*. The more Max has solved these problems to his satisfaction before entering into negotiations, the more competent Max will appear to the potential licensee. As a result, the more likely it will be that only seriously interested firms will be interested in Max's widget improvement.

POSITIONING THROUGH NEGOTIATION SKILLS

A third and final strategy involves Mad Max positioning himself through negotiation skills. Since this topic is vast, we will focus in this context on three vital aspects of negotiation: the nature of "agent type" and "agent commitment," agenda issues, and negotiation of gains versus losses. For the sake of brevity, we will utilize some of the insights of Schelling's (1960) classic, *The Strategy of Conflict*, along with the perspectives of "prospect theory" (the works of Nobel prize winner D. Kahneman and his co-author A. Tversky, for example, Kahneman and Tversky, 1979; Tversky and Fox, 1995; Tversky and Kahneman, 1974, 1992; Tversky and Shaffir, 1992a, 1992b), and the findings of the "behavioral theory of the firm" (Cyert and March, 1963), and the related "behavioral theory of negotiation" (Camerer, 1997).

The Issue of Agent Type and Agent Commitment

As has been emphasized throughout this chapter, particularly in Figure 11.3, the signals players exchange prior to negotiation are vital to negotiation success or failure. While not much research has been done in the area of applied negotiation theory as regards observation of agent "type," it is clear that type is critical here. Has Mad Max committed resources to his idea, for example by preparing a business plan to demonstrate the value of his discovery? A marketing plan? Has he mortgaged his house in order to co-finance any further development of his idea? Does his resume include other inventions either with or without patents?

Most importantly, has Mad Max investigated his opposite number's "type" before meeting her? Here, many questions need to be answered. Is Big Widget Inc. the leading producer of widget-like objects in the market? Or is Big Widget a multi-product firm currently interested in expanding its widget business? Is this the first time Big Widget has been approached by an inventor? Or has Big Widget discussed ideas repetitively with multiple inventors? One can expect a different reception from Big Widget depending on the answers to these questions. (The openness of Big Widget to other inventors is particularly important, given that the solution (5,5) in our puzzle is most likely under repetitive conditions. This being the case, Big Widget is likely to realize the self-defeating nature of the threat to steal the invention and be more willing to commit.) Finally, Big Widget may know nothing at all about the widget innovation and, trusting to Mad Max's technical expertise, consider a potential joint venture. In such a case, Big Widget will be focused on the managerial expertise of Mad Max and whether Mad Max's idea can make her "lots of lolly."

Agent type also will affect the focal and reference points in the negotiations. Sets of focal points (labels which are subliminally understood by the negotiators and can significantly aid co-operation), will differ according to agent type. Two engineers negotiating with each other will share different value sets (and implicit focal points) than two venture capitalists. Similarly, a widget manufacturer may have a different set of reference points (for example a NPV cut-off) than a venture capital firm.

A final word on commitment . . . Economics is focused on the role of "sunk costs" and the reputation mechanism as a guarantee that neither Mad Max nor Big Widget Inc. will cheat on each other. This is reflected both in the economics literature and in Schelling's observations. However, economic experiments on the role of reciprocity and altruism have revealed that if economic agents (a) have invested in a project, and (b) are mutually dependent on each other's cooperation for that project to pay-off, they are disinclined to cheat the player who commits his resources first, even though it could pay them to do so (Camerer, 1997: 170–72). If these studies are to be believed, Mad Max can be reasonably assured that Big Widget will not "take him to the cleaners" if he can be sure that these preconditions are fulfilled.

POSITIONING AND THE NEGOTIATION AGENDA

Yet negotiation success can be further assured if the negotiating players are mutually aware of the "framing" effects of the agenda and the problems of limited foresight in agreeing to a negotiation agenda.

The "game" discussed in the introduction is based on the premise that the description of the game will not affect agent behavior. In fact, economic agents are very receptive to varying descriptions of a game and alter their behavior accordingly. For example, if the game description adopted by Big Widget is essentially "we want Mad Max to establish his *bone fides*," and Max accepts this description, issues discussed will first center on Mad Max's competence or lack of same. The questions regarding the widget itself and future cooperation will be of a secondary importance, and the negotiated outcome will be affected by this.

The problems of limited foresight can be seen in experimental games of backward induction. Assume that Max wants a payoff of X, but for the purposes of negotiation increases this by an amount Y, so that what he demands is $X + Y$. He realizes that he must give up portions of Y, but feels that the negotiation will go three rounds so that he can give up $1/3*Y$ in each round, while Big Widget will do roughly the equivalent. At the end of the third round, Max suddenly realizes that the negotiation must go on for another two rounds, but, having traded away Y, the only concessions he can now give are a function of X itself. Since X reflects Max's reserve price, he must now either abandon the negotiations, or compromise on X. In either case, a lack of foresight has led to Max being placed in an uncomfortable position – with consequences for the final negotiated outcome, if any. The solution to this is to plan the negotiations, round by round, very carefully before agreeing to an agenda. This observation fits very closely with the points raised in our discussion of positioning and timing (particularly in the multiple step sequencing illustrated in Time Line Two).

Positioning and the Problem of Gains Versus Losses

While not directly related to initial signalling, the odds of Mad Max and Big Widget reaching a mutually advantageous solution are improved if they can handle the tricky issues of individual gains and losses. Over and above the problems of foresight (backwards induction) mentioned above, the significant problems here are the relationship between gain or loss and the use of a reference point, and what is termed "the social comparison of gains and losses."

Psychological studies (see for example Kahneman and Tversky, 1979), indicate that negotiators operate with implicit reference points, and that these reference points affect their perceptions of gains and losses. Assume that Mad Max previously sold another idea to a venture capital company, the payment being a lump sum of US$100 000 and a license agreement giving him a 2 percent royalty from future sales of the (his) invention. This agreement could well become a reference point for Max in his negotiations with Big Widget. Assume Big Widget is willing to offer him a US$125 000 lump sum for his idea plus a 1.8 percent royalty from future sales of his widget, indicating that the US$25 000 increase would be exactly equal to the 0.2 percent decrease in future royalty rates, in net present value discounted terms. Psychological studies indicate that Max will be more concerned about his losses than his gains, even if these equalled out! Equally important, reference points can act as "anchors" in negotiations. In such a case, Max will insist on getting nothing less than US$100 000 + 2 percent royalties. Clearly, dealing with numbers in a negotiation can be tricky for both parties.

The problem is worsened by what is known as the "social comparison of losses and gains." Negotiation agents are particularly affected by the fact that their opposite numbers simultaneously know about both their own gains, and any losses which they inflict on the agents themselves. This social comparison can lead to negotiator recalcitrance, accusations of "bad faith," and the like. To take our example of lack of foresight, suppose Max "knows" that having negotiated Y away, and having to compromise on X, Big Widget is making an amount greater than $X + Y$ herself. What Big Widget is making or not making is a function of the opportunity costs of the other money-making proposals Big Widget is entertaining. Yet despite this knowledge, and despite the fact that he is better off with a deal with Big Widget than without such a deal, Max could react against his own interests, purely because of the social comparisons he and the Big Widget negotiators are capable of making.

Given these problems, Schelling's (1960: 34–5, 44) recommendations based on examining the social processes of negotiation make good sense: Avoid numerical comparisons. Rather than focus on numbers focus on principles. Avoid discussion of gains and losses. Rather than focus on gains and losses, focus on incentives . . . Use casuistry to allow an overcommitted negotiator "to climb down" from an untenable negotiating position, thereby minimizing the social comparison effect. Clearly, if both Mad Max and Big Widget are intent on reaching a positive negotiated solution, they can do so. But negotiation is not only a science. It is an art, and can in the last instance rest on the casuistic skills and positioning of one or the other (or of both) negotiators.

THE IMPORTANCE OF POSITIONING: ROBERT KEARNS AND THE INTERMITTENT WINDSHIELD WIPER

For two decades, the lawsuits of Robert Kearns against the American and European automobile industry epitomized the struggle of our Mad Max who had been done wrong by Big Widget (the following account is based on Seabrook, 1993). The chronology of events is summarized in Table 11.1. In essence, Kearns invented an intermittent windshield wiper for automobiles in 1962, one that would vary its speed as desired, or automatically switch on or off when rain struck the automobile windshield. At the heart of his invention was an electronic control system consisting of a transistor, a capacitor and a variable resistor. The last two components worked as a timer, the transistor as a switch. The resistor would be set by the driver by a knob and would control the current flowing into the capacitor. When charged above the specified level, the capacitor would trigger the transistor, the wipers would move once. The current would drain out of the capacitor,

Table 11.1   *Chronology:* Kearns vs. Ford *re: the intermittent windshield wiper*

Date	Event
1961	Ford's principal supplier of windshield-wiper components, Trico Products, brings a new intermittent device to Ford, directly linked to and powered by the car's engine
1962	Kearns has his "flash of genius" about the intermittent windshield wiper
1963	Kearns asks Ford if they can give him some wipers to explore his idea. He develops a working model of his invention, based on new developments in electronics technology (consisting of a transistor, a capacitor, and a variable resistor). He demonstrates it to Ford engineers, who find it interesting and encourage him to continue testing it, but (later say that they believed the invention was not novel enough to be patentable)
1964	Kearns intensively tests his wiper and begins to think about patenting it. Ford does not seem overexcited about his idea, so Kearns approaches Dave Tann. Kearns assigns his patent rights to Tann's company. Tann files the first patent application in December
1967	Kearns' first patent is granted in November. Meanwhile, Kearns presents his ideas to Ford engineers and executives. According to Kearns, Ford superviser Roger Shipman promises to give him a contract, in return for explaining to Ford engineers how the wiper works – which he did
1968	Ford tells Kearns they don't want his wiper system after all, that the other engineers had designed their own. Later that year, Shipman calls Kearns to tell him his wiper might still be used for the 1969 models, and asks him to come in again and demonstrate his model
1969	Ford comes out with a new, electronic intermittent windshield wiper, the first in the industry, using a transistor, a resistor, and a capacitor in the same configuration as the one Kearns had designed
1974–7	General Motors, Chrysler, Saab, Honda, Volvo, Rolls-Royce and Mercedes, among others, put intermittent wipers on their cars
1978	Kearns, having gotten back his patent rights from Tann, files suit against Ford for patent infringement. Eventually he adds other companies to the suit
1989	By this year, Ford had sold 20.6 million cars with the intermittent wiper, and made a profit estimated at $557 million
1990	Kearns' case against Ford comes to trial. Most of Kearns' patents had expired by then. The jury finds that Kearns' patents were valid and that Ford had infringed him. Ford offers to settle the case for $30 million. Kearns turns them down. This leads to a second trial, where Kearns is awarded just $5.2 million. Kearns pitches a tent outside the courtroom and continues to push his case. Finally Kearns and Ford settle for $10.2 million
Post 1990	Kearns settles for similar sums with other auto manufacturers

and the wipers would nestle back into the hood, waiting for the capacitor to be recharged again. This invention at the time far outstripped any of the alternative designs for intermittent windshield wipers, all of which had many moving parts and various design flaws.[11]

As can be seen in Table 11.1, it was only after Kearns had attempted to sell his invention to Ford that he thought about patenting it. Kearns' initial plan, in fact, was not to license the patent rights to Ford, but to set up a manufacturing unit himself. He would thus become the sole industry supplier of intermittent windshield wipers – not only to Ford, but also to the other automobile producers. Kearns and his partner Tann applied for multiple patents on the invention. Their first patent was granted in 1967.

In 1969, Ford introduced an intermittent windshield wiper as an option for its vehicles. The wiper design was essentially the same as that covered by the Kearns' patents. Kearns sued Ford, and then Chrysler and General Motors, when the two companies introduced intermittent windshield wipers as well, based on Kearns' design. The lawsuits spread to Europe after Kearns discovered that Mercedes had also infringed his patent in 1976, and then Japan. After turning down an initial settlement offer of $50 million from Ford, Kearns eventually won a settlement of $10 million from Ford. Later he received some $20 million from Chrysler. Having fired his lawyers, Kearns missed the deadline for filing papers in his cases against General Motors, the European and the Japanese automobile firms. As a result, his claims against these companies were dismissed.

A closer look at the Kearns case demonstrates the importance of positioning. Raised in Detroit, not far from the enormous Ford Rouge plant, Kearns was impressed by the company's size and ability to make cars out of raw materials. In Detroit, one invented for the car companies, and Kearns was determined to follow suit. Partially blind in one eye, Kearns had difficulty seeing through the windshield when windshield wipers were constantly moving. He determined to find a better windshield wiper. To do this, he became acquainted with some Ford engineers working on windshield wipers, who furnished him with wiper materials on which to work. When it came time to show his prototype to Ford, he contacted an engineer, John Cupiak, who had been suggested by Kearns' brother Marty, employed in body engineering at Ford. On demonstrating his prototype, mounted in a Ford Galaxie, Kearns was informed that he had contacted the wrong man. Cupiak was a wiper linkages and blades man. Kearns should meet Ford executive engineer Joseph Neill.

Three days later, Kearns once again driving the Galaxie, reappeared for his appointment with Neill. He was surprised to find about ten Ford engineers waiting for him in the parking lot. They took turns running the wipers; they poked around under the hood; they crawled under the dash. One at a time the engineers asked him how his wiper worked. "I didn't want to tell them how I'd done it, but I didn't want to be impolite either," Kearns recalls. Eventually Neill appeared. He had a Mercury brought out of the lab and, keeping Kearns at a distance, demonstrated to him that by chance Ford was working on an intermittent windshield wiper too. Nonetheless, Neill said, Ford would like to look at Kearns' invention if Kearns would like to show it to Ford.

Neill then said he would like to know how much Kearns' wipers cost to build. He also arranged for Kearns to get instructions on Ford's specifications test. Kearns left in a state of euphoria (Seabrook, 1993: 5).

Analysed from a positioning angle, it can be seen that there were several flaws in Kearns' initial contacts with the Ford Motor Company. For his early experiments, he accepted Ford wipers as gifts, which would naturally lead Ford to believe he would give something back in return. For his first meeting, he utilized his brother's information as to whom to talk with, which turned out to be wrong; Kearns would have had greater credibility had he gone to executive engineer Neill directly. In his meeting with Neill, he let Neill set the agenda. The lack of reciprocity was an ominous sign: Ford engineers evinced interest in Kearns' invention, thoroughly inspected his Galaxie, and quizzed him about the details of his invention – but did not allow Kearns to inspect their own Mercury. From a positioning point of view, after this experience, Kearns might have been better served by approaching some alternative customers, either other automobile companies or their suppliers, such as Ford's windshield wiper supplier Trico. Would such additional contacts, or the mere mention of his planning to make them, have changed the situation? While it is hard to say, there is little evidence that this tactic would have hurt his standing with Neill.

Ford requested Kearns to subject one of his prototypes to their specification test. But before agreeing to do so, Kearns did not request any agreement from Ford governing the future use by Ford of the information gained. After Kearns' prototype fulfilled the demands of the Ford test, Kearns attempted to contact Ford. Ford appeared lukewarm to any further cooperation. Kearns still had to file his patents, and was pressed for finances. He approached David Tann, an owner-manager (together with six brothers) of Tann Corporation, a midsized supplier to the automobile industry. Tann was enthusiastic, and Kearns and Tann agreed that Kearns would assign his rights to Tann, who would cover patenting costs. Furthermore Tann would pay Kearns $1000 a month to cover the costs of his wiper R&D program, plus royalties when the wiper went into production.

Tann was better at positioning than Kearns. Having applied for the patents, he contacted Ford. Tann's and Kearns' demonstrations made to Ford engineers and executives were well received.

Finally, Roger Shipman, a Ford supervisor, announced to Kearns that he had "won the wiper competition." He told Kearns that his wiper would be used on the 1969 Mercury line. Kearns was given the prototype of a windshield wiper motor to celebrate the occasion. The other engineers welcomed him aboard Ford's wiper team. Then, according to Kearns, Shipman asked him to show his wiper control to the rest of the team. Wipers were a safety item, Shipman explained, and the law required disclosure of all the engineering before Ford could give Kearns a contract. This sounded reasonable to Kearns so he explained to the Ford engineers exactly how his intermittent wiper worked (Seabrook, 1993, p. 11).

Kearns thus disclosed his secret know-how without receiving any contract with Ford. Five months later, Ford informed Kearns that their engineers had designed their own wiper system and that his wipers were not wanted by Ford at all. In 1969, Ford marketed its first intermittent windshield wiper, based on the transistor, variable receptor, and capacitor described in Kearns' (now Tann's) patents. Lacking a contract with Ford, or even any kind of informal understanding (which might have led Ford to compensate Kearns in some fashion after all), Kearns now had no choice but to withdraw from the fight, or sue.

The degree to which Kearns' positioning strategy was affected by his lack of resources is unclear. What is striking is that his positioning was disastrous. He should have known that one of Tann's major customers was the Ford Motor Company, and that as a result, Tann could not afford to join him in suing Ford for patent infringement, should that become necessary. Kearns later had to sue Tann in order to get his patent rights back. Similarly, by this time, Kearns should have been aware that Ford was not entirely to be trusted. He might also have done some research on the company, and found out that Ford in principle disliked patents (Henry Ford had claimed they blocked innovation much more than they encouraged it!). Some clever positioning could have given him a much stronger negotiating stance. He could either have revealed his secret know-how about the invention to Ford engineers on a *quid pro quo* basis (positioning through timing), or involved his own lawyers in the Ford contract negotiations (positioning with multiple agents). He might perhaps have offered to sell the patent rights to Ford. Kearns resorted to none of these methods. It is not surprising that the Ford engineers lost all respect for Kearns. (Kearns recalls that in his last meeting with the Ford wiper team, one of the engineers taunted him as he was leaving.) Throughout this story, as can be seen in Table 11.1, there was thus a marked lack of reciprocity on both sides. One result was that Kearns and Ford failed to build up any real good will in their relationship, which could have led to the implementation of another kind of solution that benefited both sides.

Another weakness in Kearns' positioning is that he was less interested in making money out of licensing (or selling the patent rights) than he was in setting up his own company and selling intermittent wipers not only to Ford, but to all automobile companies using his invention – a strategy which automobile industry veterans undoubtedly advised against. Nonetheless, one of the major points of his subsequent lawsuits was a demand that all infringers of his patents should be forced to stop manufacturing intermittent windshield wipers and to switch to his supply company for their intermittent windshield wiper needs, a claim that no court would have enforced.[12]

CONCLUSION: MAD MAX AND POSITIONING REVISITED

We began this chapter with a fable, the genius inventor (Mad Max) whose ideas are ruthlessly exploited by a capitalist firm (Big Widget Inc.). This fable exists in various versions throughout the western world. Ideas can be appropriated. Inventors can be cheated. There are problems with the information as a fungible commodity.

While we do not wish to be thought unsympathetic to creative inventors, our findings leave us with the following thoughts:

- first, many inventions may not have any commercial potential to begin with, irrespective of what the "mad" inventor may have thought;
- second, although inventors are "ripped off," this may be more a reflection of their lack of positioning skills than of any major wrong-doing on the part of their financial partners;
- And, third, those inventors who have been "ripped off," as was the case with Kearns, might have avoided this fate by observing the signals of their opposite number more carefully, positioning themselves according and by more carefully choosing partners with whom to negotiate. In short, the problems of the "ripped off inventor" are quite likely exaggerated.
- Fourthly, sources in the windshield wiper business have informed us that Kearns' protracted patent battles created many problems for inventors in other companies who sought to patent improvements of the intermittent windshield. Their employers advised them not to apply for patents until the Kearns' case had run its course, to avoid exposing themselves the possibility of further lawsuits (and the associated bad publicity)!

What is notable is that an examination of the negotiation processes surrounding invention and finance can yield a wealth of information so that the numbers of inventors who have been "ripped off" declines significantly, both through a better realization of the purchaser/potential sponsor's point of view and through a increased ability to negotiate terms for themselves.

What is true of our inventor is equally true of his negotiating counterpart. To be successful Big Widget must be cognizant of the particular problems of the "idea salesman." A lack of such cognition is not only a sign of a poor businessman, but can lead to a poorer society as well.

NOTES

1. The linkage between positioning and strategic reflexivity has also been made in relation to bidding contests for telecom licenses, see J.D. Davis (2001).
2. Thus Big Widgets payoffs from stealing are 8.5 plus 1 (learning experience), from cooperation 4 plus 1 (learning experience), and Mad Max's are 4.5 plus 0.5 (learning experience) in the case of cooperation. Note that the pay-offs are different from the actual payments between the two parties. Thus, Mad Max is anticipating 4.5 payment, not 5, from Big Widget in the case of cooperation as 0.5 is for learning experience and is a different form of payoff. The same logic applies to Big Widget.
3. We have given this payoff a value of 1 because we assume that even if Mad Max does not reveal his secret know-how, Big Widget will still learn something valuable from the negotiation. For example, she will learn more about the background for Max's invention, additional information related to the invention that is not secret know-how but which is still important in attempting to gauge the chances of its commercial success, and a sense of what kind of inventor and negotiator Max is.
4. This is in accordance with our earlier logic where we have separated the values of learning from the negotiation from the other payoffs.
5. Note that Mad Max earns 0.5 through the learning experience and this does not enter into the question of his compensation which will be net of 0.5, or 2.25.
6. In Åstebro's study (1999), as cited in the Introduction, the probability that an invention by an independent inventor would actually reach the market was even less than this (6.5 percent).
7. For example, Big Widget could engage in subtle signaling, evincing a clear interest in Max's invention, of asking Max to develop a prototype that fits her specifications. This process is part of the larger negotiation game where the parties size each other up and "feel each other out." She could also commit resources to the project (in a limited manner), for example by asking her engineers to specify what the invention must do to be commercially promising, asking her lawyers to draw up a draft of a contract with Mad Max, or asking the marketing department for a preliminary evaluation of the commercial prospects of the invention.
8. For reasons of space it is not possible to pursue all variations of the question of timing. One important aspect in this regard is the decision as to when to patent the invention. The patent law requires disclosure of the patent document 18 months after the Patent Office receives the patent application. But during this 18 month period, the inventor retains priority to the invention; no one else can apply for and receive a patent for the same invention. As a result, it might make sense for Mad Max to time his initial approach after applying (and thereby with priority to the invention) but before the 18 month period is over. (It should be noted that in the case of Robert Kearns, as described in the next section, the then-current US patent law did not require disclosure after 18 months.)

9. By purchasing a license option, the firm reserves the chance to enter into a license contract in the future.
10. It should be noted that many invention service firms do engage in fraudulent behavior. For instructions as to how to select reliable invention services firms see US Federal Trade Commission, "Invention Promotion Firms," at www.ftc,gov/bcp/conline/pubs/services/ invention.htm (accessed 8 February 2007).
11. For example, Trico Products, Ford Motor Corps' major windshield wiper supplier developed an intermittent windshield wiper which had 29 parts, failed when the automobile accelerated, and promised much trouble and many repairs should it have been adopted.
12. Or in the terms of the court decision in one of the lawsuits brought by Kearns: "entitlement to an injunction implementing the right to exclude, as compared to assessing damages against an infringer, is not absolute, even during the life of a patent but is discretionary." Ford and the automobile firms have claimed that Kearns' patents were invalid because they used obvious technology, and that they were in the natural evolution of technology at the time. This should not surprise overly much as many patents involving electronic components have been ruled invalid on similar grounds.

REFERENCES

Åstebro, T.B. (1998), "Basic statistics on the success rate and profits for independent inventors," *Entrepreneurship Theory and Practice*, **23** (Winter), 41–8.

Åstebro, T.B. and K.B. Dahlin (2005), "Opportunity knocks," *Research Policy*, **34** (9), 1404–18.

Bazerman, M.H. and M. Neale (1992), *Negotiating Rationally*, New York: The Free Press.

Camerer, C.F. (1997), "Progress in behavioral game theory," *Journal of Economic Perspectives*, **11** (4), (Fall), 167–88.

Cyert, R.M. and J.G. March (1963), *A Behavioral Theory of the Firm*, Englewood Cliffs, NJ: Prentice-Hall.

Dahlin, K., M. Taylor and M. Fichman (2004), "Today's Edisons or weekend hobbyists: technical merit and success of inventions by independent inventors," *Research Policy*, **33** (8), 1167–83.

Davis, J.D. (2001), "Bidding for telecom licenses strategic reflexivity and the 'winner's curse,'" INT Institute Seminar, Handelshøjskole i København, Copenhagen, 24 January, 32pp.

Davis, L.N. (2004), "Intellectual property rights, strategy and policy," *Economics of Innovation and New Technology*, **13** (5), 399–415.

Kahneman, D. and A. Tversky (1979), "Prospect theory: an analysis of decision under risk," *Econometrica*, **47** (2), 263–91.

Mazzoleni, R. and R.R. Nelson (1998), "Economic theories about the benefits and costs of patents," *Journal of Economic Issues*, **32** (4) (December), 1031–52.

Parker, R.S. and G.G. Udell (1996), "The new independent inventor: implications for corporate policy," *Review of Business*, **17** (3), 7–11.

Schelling, T. (1960), *The Strategy of Conflict*, Cambridge, MA: Harvard University Press.

Seabrook, J. (1993), "The flash of genius," *The New Yorker*, 11 January, accessed 26 February, 2007, at www.booknoise.net/johnseabrook/stories/technology/ flash/index.html.

Tversky, A. and C.R. Fox (1995), "Weighing risk and uncertainty," *Psychological Review*, **102** (2), 269–83.

Tversky, A. and D. Kahnemann (1974), "Judgement under uncertainty: heuristics and biases," *Science*, (185), 1124–31.

Tversky, A. and D. Kahnemann (1992), "Advances in prospect theory: cumulative representation of uncertainty," *Journal of Risk and Uncertainty*, **5** (2), 297–323.

Tversky, A. and E. Shaffir (1992a), "The disjunction effect in choice under uncertainty," *Psychological Science*, **3** (5), 305–9.

Tversky, A. and E. Shaffir (1992b), "Choice under conflict: the dynamics of deferred decisions," *Psychological Science*, **3** (6), 358–61.

Udell, G.G. (1990), "It's still caveat, inventor," *Journal of Product Innovation Management*, **7** (3), 230–43.

PART 4

Sensemaking

12. Sense caring in innovation

Peter Hagedorn-Rasmussen

Innovation is a process of collective sensemaking. Successful entrepreneurs are those who are able to create and sustain networks by careful attention to a process of ongoing sensemaking. This is an emerging and unfolding process which may be targeted toward building a sustainable business concept. Although innovation processes are characterized by a high degree of arbitrariness, they require a great ability on behalf of central actors to translate the emerging business concept as it unfolds. It must fulfill the needs of the anticipated customer; it must fulfill the legal requirements and so on. In the process of translation, meaning is negotiated around the unfolding artifact – and knowledge is embedded in the innovation. This chapter explores how care plays a role in creating sense within a net of actors engaging in entrepreneurial activities.

The chapter draws on a case-study which explored the origin of an e-based realtor service.[1] The original idea of the entrepreneurs was to develop a realtor marketplace on the internet – a place for sellers and buyers to meet – and a place where profits were moved from the realtors to the users. It was based on a philosophy of a large "E." However, as will be illustrated, the business concept unfolds, and, as a result, the innovation is translated. The original idea of the large "E" becomes a little "e." The Internet is still central to the business processes, however the business concept has moved closer to that of the traditional realtors – the business concept competes *within* the realtor business. In part, this is a simple story that illustrates successful adaptation toward the market conditions and the user's e-confidentiality and so on. However, looking closer, we can learn something about *how* sense is created and how it is cared for within the net of actors. We may – as a result – achieve a better understanding of the unfolding character of innovations.

The chapter explores two approaches to the meaning of care in innovation: First networking is increasingly important in processes of innovation. The strength of a network is dependent on the ability to create a process of alignment between the business concept and the human resources and capabilities that are enrolled in the network. This requires a caring approach to managing an unfolding net of actors, where each actor brings unique resources and capabilities to a not yet fully realized artifact. It is not about strategic planning. Strategic planning requires a well thought out concept concerning the particular innovation. Instead it is about strategizing with care – caring about the environmental cues that must be reflected upon and continually embedded in the unfolding innovation. Second, in the process of creating a sustainable business concept, collective sensemaking is a critical factor. This implies that relevant agents are sensitive to the environmental cues which they scan and notice. Sensemaking might be of experienced market disequilibria, new market opportunities, promising technological opportunities or human resources that provide new capabilities. The relevant agents have to be able to enrol these into the emerging net of actors and to translate them in the further development of the business concept. One challenge is that most sensemaking is said to be made retrospectively, while strategizing implies an imagined future direction. In this chapter, I explore how these two seemingly uneasy pairs – strategy and sensemaking – intersect when entrepreneurs care about an unfolding innovation.

Throughout the chapter, I refer to ICT-based[2] knowledge services, both in general and in practice, as innovation. However, sometimes I also use the terms "business concept" or "business model" to refer to the innovations. This latter use emphasizes the contextual character of the innovation. It highlights the role of the customers, and the strategic requirements which must be reflected in an innovation in order to make it sustainable in the market. The case studied here is taken from the private sector. However, the theoretical perspectives may be applicable in the public sector as well as in innovation which results from partnerships between distinct fields. The role of the customer (citizen), and the role of the market (the user, the "common good") are the environmental cues that transform the net of actors, and these would be quite different. However, successful innovations, irrespective of the field, are characterized by their ability to create a sustainable network where collective sense is created and sustained as the network and the innovation emerge.

The chapter begins by elaborating two theoretical perspectives, emphasizing nets of actors and collective sensemaking. Following this, is a section which presents a narrative version of the case, the e-based realtor business RobinHus. It introduces RobinHus and describes episodes of the innovation process that make it possible to explore the themes presented in the theoretical perspectives. Finally, I summarize some of the main findings related to the interrelationship between care, networking and sensemaking.

INNOVATION AS EMERGING NETWORKS

The understanding of micro-sociological processes may be promising for understanding new dimensions of the innovation processes. When nets of actors and networks emerge as a result of arbitrary as well as intended endeavors something new is created – but what are the practices and the relationships actually doing? How do they relate to each other?

Theoretically, the idea of exploring nets of actors reflects an interpretative approach (Burell and Morgan, 1979) in an area mainly dominated by functionalistic approaches (Jennings *et al.*, 2005). The approach emphasizes the relational aspect of innovation.[3] Empirically, the concept of networks seems to be of increasing importance in processes of innovation (Sundbo, 1998; Sundbo and Gallouj, 2000) as well as processes of institutionalized organization (Snow and Miles, 1992). The strength of a network is dependent on the ability to adapt and transform itself – and the business concept – in the emergent process of analysing strategic opportunities and threats.

While entrepreneurship is often ascribed to individuals that have great aspirations and are motivated by desires to create and realize a vision by creative destruction (Schumpeter, 1975 [1942]) or to conquer disequilibrium at the market (Kirzner, 1992), I find that neither of these can stand alone, however none of these can be missing. Both are necessary for entrepreneurship to occur. Both have to be enacted and followed by relevant actors within a network that realizes a specific innovation. The network is nothing without the entrepreneurs – however the unfolding entrepreneurial activities cannot be sustained without the collective net agents. They apply more of their resources, in creating and establishing sensemaking, than merely realizing the instrumental knowledge they have appropriated before.

Building a network that may produce a sustainable business concept is also about the strategic management of human resources and capabilities (Hamel and Prahalad, 1994). While the strategic management of human resources and capabilities in general is seen as a critical factor in the knowledge economy, this should be particularly relevant for nets of actors and organizations that innovate within (e-based) knowledge services (Lengnick-Hall and Lengnick-Hall, 2003; Wright *et al.*, 2005; Salaman *et al.*, 2005). A process of innovation depends on the network's ability to bridge the emerging business concept – as a result of analysing the opportunities and threats – with the (human) resources that are enrolled and retained in the network. This is often reflected in visions and strategic business plans for developing innovative service packages. However, we know that the practical processes are often emergent and conditioned by our restricted capability to gather full and accurate information. The most intriguing challenge is that we can never be aware of that little piece of important information which we do not know, before we have realized that we did not know it – and therefore we base our actions on an inaccurate foundation. We act within the limits of bounded rationality.[4]

Therefore it is of decisive importance to combine strategy making with an ability of the network which is responsive and reflexive. There is a need to acknowledge the valuable dimensions of serendipity in processes of innovation. Serendipity, explored from a sociological perspective by Merton and Barber (2004), is the phenomenon of finding valuable things not sought for. It is the faculty of making happy and unexpected discoveries by accident. The sustainability of the network in the emergent process is dependent on the ability to create collective sense around the strategy and the business concept. This includes the ability to be responsive to discoveries that were not sought after from the start, but which emerged by accident in the process. Sense is the glue in the network. In particular, the loosely coupled agents would dissociate themselves if sense were lost. The network would either ossify or evaporate.

TAKING CARE OF COLLECTIVE SENSEMAKING

The second perspective thus emphasizes that innovation processes are really processes of continual *sensemaking* (Weick, 1979, 1995). Which cues are scanned, noticed and enacted by the relevant actors and actor groups is decisive for how a process and emerging business concept unfolds (Dougherty *et al.*, 2000).

Weick has proposed seven properties of sensemaking which are important in exploring processes of organizing (1995), which I will briefly relate to the process of innovation. *First* sensemaking is about *identity construction*. Individuals who engage in innovation processes engage in realizing desires – on the basis of the past experience, knowledge repertoire and social capital, they explore opportunities to be realized. Entrepreneurs may have localized market disequilibria, technological opportunities and so on which they imagine they are able to utilize by realizing an entrepreneurial idea. These appear as cues in the environment, and some are enacted later in the process. In the late 1990s, the Internet sprang up as an opportunity for creating business concepts. As we shall see, this cue was enacted by entrepreneurs who created nets of collective action by merging the opportunity with cues such as apparent market disequilibria on the real estate market. Thus the realization of the business concept is dependent on the entrepreneur's ability to engage in collective nets that provide different resources (financial, knowledge, technological and so on). These resources form synergies which help to realize the opportunities on the one hand, and defeat the threats and weaknesses on the other. Even though sensemaking is said to be retrospective, I suggest that the practice of building such networks reflects a strategic orientation where entrepreneurs care about building a net of collective identity around the imagined innovation. The strategic orientation is important as a "glue" in the collective net since it represents the belief in the opportunities of the imagined innovation – attracting venture capital for instance rests upon the ability to present a strategic plan which has a convincing business concept. At the same time though, the innovation is the unfolding artifact which reflects the entrepreneur's desires. It is unfolding precisely because the collective net needs to translate the imagined innovation. This would be in response to the conditions (technological, financial, human resources and so on) which continually challenge the innovation. Thus, a successful innovation is dependent on the net for continually building a sort of collective identity – a sense of direction – as well.

Second since the world is filled with too many meanings, equivocality, sense is made retrospectively. This is a very delicate matter when we want to explore strategizing processes and to explore the process of how entrepreneurs realize – and continually translate – the ideas they have. As we saw above, entrepreneurship is always based on a fundamental belief and desire to achieve the realization of an imagined artifact, the innovation. This requires a sense of direction – a plan. Most often we assume that actors act in accordance with the reasons they give: they get an idea, transform it into a goal and make a plan to fulfill the goal. Then they act in accordance. However, the retrospective aspect of sensemaking suggests that actors more commonly (en)act, and then reason. The realization of the e-based realtor business, which I will return to, is partly fueled by the belief in a sustainable business model which, financially, rests upon banner advertising. This belief in a rationalized business model is fundamental for the entrepreneurs to engage in the process of innovation: If this had not been the case, the entrepreneurs may never have started. However, this financial model appeared not to be sustainable. The entrepreneurs had to translate their imagined business model. They had to enact other cues and translate their business model and their innovation as a consequence. The concept care, I suggest, represents the ongoing interplay between the strategic "plan" – a rationalized representation of the innovation – and the retrospective nature of its unfolding realization.

Enactment thus occurs as a result of the interaction with the sensible environment – the third property of sensemaking (Weick, 1995). Innovation occurs as a process, where individuals and nets of actors engage in sensemaking processes produced as a result of continual interaction. This to some degree resembles the framework suggested by Bijker (1997), which has a heritage in symbolic interactionism, where technological frames are established as a result of recursive processes between relevant social groups and the technological artifact in question.

The fourth and fifth propositions of sensemaking concern sensemaking being social and ongoing. These propositions are clearly related to my first theoretical perspective, namely that understanding innovation processes requires sensitivity towards nets of actors and the unfolding character of innovations.

As a sixth proposition, Weick suggests that sensemaking is focused on extracted cues. Because of equivocality, individuals and nets need to focus on some cues, making others peripheral. People extract cues and "take faith" in them as a means of achieving sensemaking. The rationalized business model, mentioned above, represents a means of sensemaking, where the financial model rests on the banner advertisement. The entrepreneurs put an effort into extracting cues which may provide sense to the unfolding innovation – they intentionally extract cues in order to sustain a process of collective sensemaking, thus fueling the process of creating an e-realtor business. If this does not succeed, the network will be dispersed and the potential synergies within the collective network will get lost. Which cues are extracted is decisive for the unfolding and emerging innovation. In the process, the enactment of cues may result in canonical practices that institutionalize the basis of a successful innovation: a strategic vision, an organizational structure, a technological artifact, a piece of software, a marketing program, a relevant customer group or an e-based realtor service.

This is said to be driven by plausibility rather than accuracy – the seventh property of sensemaking. I will not delve into this, but merely quote Weick: "that accuracy is nice, but not necessary" (Weick, 1995: 56).

Getting into the Field

It is probably neither new nor surprising that to realize a vision more resources are necessary than those provided by the agent who got the idea. However, entrepreneurial research sometimes emphasizes the "entrepreneurial prince," thus the research becomes a self-fulfilling prophecy about the qualitative character of the entrepreneur. This is why I insist on following explanation of how networks emerge – to explore the character of the network. History is written backwards – as is most sensemaking. Looking at an innovation and the entrepreneurial idea that made this come into being challenges our intellectual ability to understand how the network actually has been able to succeed. The reason is that we often explore the narrative of the "winners," since it is the innovations that have succeeded which we are able to study. As scholars, we follow the prince and may challenge our bias by not only probing into how it – the phenomenon we explore – came into being, but also why it might be that the innovation did not become something else. Looking at whatever issue – organizing, strategizing, innovating – as practice is a difficult path, because we must acknowledge that the description of these episodes in detail – as everyday practice – may, in the first instance, look banal. It challenges our desires as scholars to create something generally applicable. We may renounce generating general applicable knowledge as our task, and make ethnographically inspired thick descriptions which provide silent actors with a voice. However, I do not renounce the task of generating knowledge that is generally applicable. Instead I insist that we need to dare to delve into practice and the seemingly banal. I have to trust that "in there", in the tacitness of everyday practice, we as scholars, who for most of our time engage in rather abstract discussions, can find something valuable and unique. This is why I engage in presenting a narrative version of the case-study in this chapter.[5]

CREATING AN E-REALTOR BUSINESS – ENTREPRENEURIAL DESIRES

RobinHus – in English "RobinHouse" – was a web based real estate agency launched in 1999 by two young entrepreneurs. Torben and Frederik had known each other for a long time, and often discussed the opportunity to create "something" lasting as entrepreneurs. Their desire was to create something from the ground. Their desire to create something was however neither based on a specific idea, nor was it directed toward a particular market. Their first aspirations were formed by cues related to the hype surrounding the Internet. Their aspirations concerning the Internet and information technology – although not necessarily focused on entrepreneurial notions – had formed the trajectories of their formal education.

Besides this, both were brought up in families which had traditions of having independent businessmen. Torben had recently finished a degree from a business school in the Management of Technology, specializing in innovation in knowledge intensive businesses. Frederik likewise had finished his formal education within the Law, writing a master's thesis on "The electronic presentation of digital information services." Furthermore, he had been traveling and had achieved a degree as an haute couture designer in Paris and Milan. After their formal education was finished, Torben and Frederik quickly found positions within each of their areas. Torben took a position as a "controller" within a large financial institution. His central task was to evaluate new business concepts that were suggested to the board. Frederik took a position as a legal counsel on an Internet service called Netdoctor which specializes in providing knowledge on health issues.

It was the heyday of the dot.com's. The opportunities of creating lasting innovations seemed to be endless. The optimism was high. If everything went well, although this was not the only criterion for success, they would be able to sell the concept later – harvesting a large revenue.

BOX 12.1   THE PRODUCTS OFFERED – AND A TASTE OF THE UNFOLDING INNOVATION

The first type of advertisement mentioned below is the one closest to the original idea. The second and third reflect the incremental innovations which have taken place as a result of the emerging business concept. The last is the one which makes up most of RobinHus' turnover today.

- *"The Start Advertisement"* includes a picture of the house, a plan (drawing) and a description. This does not give a direct match with the buyer card index. It is for free.
- *"The Pro Advertisement"* includes information given in the Start Advertisement, however it also provides the seller with the opportunity of uploading more pictures, slideshows and so on. Furthermore, the seller has access to discounts on newspaper advertisements. In the system, the seller fills out an estate agent's information sheet. As such the seller provides the same information to potential buyers, as traditional estate agent services, would do. As a customer, the seller needs to provide the information, which may create a greater risk of failures or misinformation. Most importantly the seller has access to the buyer's card index which automatically matches the seller's real estate with potential buyers and automatically sends them an e-mail. The price for this service is a flat rate of around 133 euros.
- *"The RobinHus Estate Agent"* service is closer to the traditional estate agent. In a network of estate agents that are associated with the back office, sellers and buyers receive nearly the same service as they would receive at traditional estate agents. The agent sets up the necessary information sheets and so on. The difference is that the seller shows his or her real estate to potential buyers. The price was – very unusual for the field – a flat rate price of 3351 euros of which 670 euros had to be paid up front.

Utilizing Disequilibrium in the Real Estate Market – the Conception of an Idea

The entrepreneurial idea was conceived as Torben was about to sell his apartment. Serendipity plays a role, since neither of the entrepreneurs was focusing on real estate markets. They sought, more generally, after opportunities to realize their desires for becoming independent. In this process they made an unexpected discovery by accident. Torben's experience was that the costs associated with selling the apartment were way beyond the service associated with the transactions which were provided. While having no professional knowledge about the real estate business, Torben and Frederik discussed the "market disequilibrium." In their judgment, the internet provided opportunities for creating a market-platform – a virtual space – in which buyers and sellers of real estate could meet and make the necessary transactions making traditional real estate agents superfluous. The required legal services relating to real estate transactions could be provided by lawyers who would be associated as partners on the website. While Torben and Frederik did not have knowledge about the realtor business, they did have access to it. Close relatives had been engaged with estate transactions for many years as lawyers. This provided them with important capabilities and resources in terms of building a sustainable network of actors supporting the unfolding business concept. But even more important, it provided them with insights for seeing where the opportunities and the limits for codifying knowledge were. They named their business "RobinHus" (RobinHouse), which had connotations with the folklore hero Robin Hood who as we all know was eager to conquer the disequilibrium of distributive justice. This connotation was also given a practical meaning since RobinHus offered the service for free.[6]

Keep it Simple – Let the Internet do the Thing

The platform Torben and Frederik began with was very simple, thus keeping the costs and financial requirements low. A publicly accessible website, the knowledge service per se, and a simple internal dashboard where tasks were summarized in order to be prioritized and later implemented. At the time, all the content was developed by Torben and Frederik: the description of content, design of the website and marketing material. They worked to increase users' accessibility in search engines for instance. The programming/coding of the website was outsourced, purchased at an IT consultancy. A transaction that the two entrepreneurs felt was one of their small initial successes, since they were able to acquire it for at fixed, negotiated, low price. They made a contract with all the necessary requirement specifications and

---

BOX 12.2   THE E-REALTORS' FOUR IT PLATFORMS

1. The website itself – for (potential) buyers and sellers.
2. "My Advertisement" – for sellers. Here the customers that want to sell their estate have a work bench, where they can set up and improve their advertisement. The seller may create a slideshow, set up new Open House Arrangement, order newspaper advertisements (extra), follow statistics on how many have visited the site and so on.
3. The RobinHus Internal Dashboard. Practically everything can be controlled from here. Advertisements, follow up calls for customers, standard e-mails (for instance if necessary to correct an incorrect use of the different types of advertisements), control of material that is sent to customers (advertisement material, signs to put in front of houses and so on).
4. RobinHus Estate Dashboard – the dashboard from where the local estate agents operate together – with their customer's sites.

---

that provided them with a low price but high quality (relative to their wishes at the time) website.

The entrepreneurs had confidence in their idea and expected the Internet to provide the opportunities for accessing a large customer base. In the first phase, the entrepreneurs participated in a network of entrepreneurs, *First Tuesday*, where the members' only common interest was to utilize the opportunities of the Internet. The network, however, I suggest, fueled the entrepreneurial desires and hopes, supporting the collective process of sensemaking. Although having a minor role, the network at the time did perform a kind of glue for the business concept. It supported the experience of trust in the sustainability of the business concept. The confidence also included a belief that the prospectively growing customer base would provide them with support to attract further financial opportunities through venture capital. A large customer base would supposedly also make the site attractive thus creating an opportunity for selling banners on the website to relevant businesses, associated with real estate transactions (furniture removers, architects, interior designers, carpenters, plumbers and so on).

In this first phase, RobinHus was driven by the two entrepreneurs with an organic expansion of the network: Relatives with professional experience within the realtor business, restricted investment in IT-programming, building up business relations with advertising agencies and so on. After this preliminary phase was finished, Torben and Frederik hired a student on an hourly basis to do the telemarketing. Later he became a fulltime customer service manager. They also hired an IT consultant to develop the website. The person only had to make the necessary descriptions and they outsourced the programming, however, later on, they insourced the programming as a means to control the incremental transformation of the site.

Meeting the Wall – How Does One Codify Trust?

The two entrepreneurs did not have to wait long to recognize that they may have overstated the ability of the idea itself to attract customers, venture capitalists and advertising clients. They experienced how many of their kindred spirits within First Tuesday were declared insolvent and went bankrupt as a result of the dot.com recession. While the name RobinHus fitted well in the beginning and provided them with a large amount of good press, being a welcomed challenge to the traditional realtor business, they also found that the coin had a flip side, which had less to do with the brand name but more to do with the business concept: Lack of trust. This in fact underscores the challenge of codifying knowledge. If we acknowledge that the practical knowledge associated with real estate transactions also implies the tacit dimension of establishing a trust relationship, we can better understand why this e-realtor goes through a process which can be characterized as a process from being a big "E" to little "e," from only "clicks" to embracing "mortar" represented by the counselling provided by realtors. An anecdotal example was given by the entrepreneurs whereas friends that had praised their initiative, when it came to their own transactions, chose traditional realtors. They did not really trust the internet as the best means of engaging in a transaction that only occurs a few times in a lifetime. The service provided in face-to-face interactions with a traditional realtor, and the tangible paperwork associated with it increased the experience of trust in the deal, thus giving the customer a sense of certainty and security in the process of making a deal.

As a result, the entrepreneurs chose to translate the business concept. They started building up a network of local realtors who were associated with RobinHus as independent realtors, meanwhile taking degrees as realtors themselves. As a consequence, they began to provide services that were much closer to the traditional realtors (see Box 12.2).

Organizational Sensemaking Within the Newly Established Network

RobinHus enrolled a net of independent realtors who worked in their local areas. They provided estate agents, previously working for competitors, with an opportunity of achieving a high degree of independence, but also a brand and a "spinal cord" represented by the ICT platforms. Even though the income for the individual estate agent was, in many cases, lower than that offered by competitors, RobinHus did not meet any difficulties in attracting them. The reasons for this, as expressed by one of the estate agents we interviewed, was the possibility of achieving higher flexibility and independence. Also, the real estate agents expressed an experience of contributing to reducing the market disequilibrium, which the interviewee saw from a moral perspective, and which I interpret as a means of identity construction that was supported by the RobinHus (Robin Hood) brand name. This actually characterized the overall culture in RobinHus, where several employees expressed the feeling of serving a greater purpose in providing this – in their interpretation – unique low price high quality service. However, the brand name also had a flip side, since the connotation of Robin Hood, although the folklore implies a noble and underlying greater good, after all has to do with theft. As a result the name was no longer used to position the company, as it was in the first years, as the impertinent competitor within the overall realtor market. On the contrary, the company in the most recent phase, up until the case study was undertaken, emphasized their role as a serious and equal competitor on the market. This, for instance, was supported by a more direct marketing strategy in the traditional realtor advertising spaces: newspaper advertising. Whereas previously, it was considered that the Internet was the only necessary and important place to appear, the strategy now was to be considered as a more serious and "mature" – less impertinent – competitor by appearing in the traditional landscape of realtor advertising.

PRACTICING INNOVATION – EXPLORING THE OPPORTUNITIES OF THE NET

> We are neither inventors nor geniuses. We are simply utilizing the opportunities that exist right in front of us all.

This is how Frederik and Torben reflect upon their innovation.[7] It emphasizes the perspective of strate*gizing*, rather than strategic planning. It emphasizes how networks emerge as a result of an emerging practice where cues are extracted and noticed (it exists in front of us all), but also applied in practice (we utilize it). This includes their own reflection on how their role has been to manage a net of resources consisting of human resources, market and technological opportunities and so on. The entrepreneurs began their journey while the dot.com hype surged – in 1998. Today they have created a business that is quite different from the visions and intentions they had in 1998. Today eight people work at the office in Copenhagen to provide the Internet service. Despite the initial ideas of being primarily a meeting place for buyers and sellers, they now also have a loosely coupled network of more than 20 independent realtors in different areas of Denmark.

Driven by the desire of "creating" something new, the two entrepreneurs were scanning the environment for "holes" in which they could create new paths based on the opportunities within ICT-services. Their desire was not to primarily do "big business," but to seek out opportunities for reducing the transaction costs, by creating an ICT-based platform where sellers and buyers could meet. Without any educational background in realtor business, but with degrees within the management of technology and law as well as design, the entrepreneurs began their journey to try out new paths. As such it was a crusade against what they found was a market failure. The innovative journey has been a process characterized more as one of practicing "trial and error," an ongoing process of scanning the environment for resources or opportunities to solve practical problems as they arose, than one of "design and planning." In the process, however, their original ideas as a result of unresolved opportunities, have been transformed. They have created a more traditional organizational base for the further development of the business concept.

Interesting though, is that the realization of the e-realtor service is not based on the entrepreneurs' ability to focus on their initial vision, and blindly follow this by requiring accuracy in the way they build up knowledge resources and capabilities. The realization is a consequence of their ability to adapt, translate and "revise" the business concept. The concept and idea are like amoebas that unfold as they meet new barriers and challenges – an amoeba unfolds as a result of successes and opportunities that support the direction and makes enactment possible.

CONCLUDING REMARKS

In this chapter I have set out to explore two different, but also overlapping perspectives relevant to care in innovation. First, it is crucial, in processes of innovation, to create and sustain networks of critical capabilities and resources. It may be argued that in exploring innovations which have demonstrated sustainability, this will be a self-fulfilling prophecy. "Critical capabilities" and "critical resources" are abstractions that, by definition, distinguish successful strategies from unsuccessful strategies, and successful innovations from unsuccessful innovations. This perspective also includes the statement that successful networks are those which are able to adjust, and enrol critical resources, as a consequence of an emerging strategy. The statement may be rendered guilty of the same objection, but when we set out to explore the translation of the emerging business concept we may acknowledge the interrelatedness between the internal strategy, the business concept, the specific innovation on one hand and, on the other hand, the external opportunities and threats which mutually transform each other. The entrepreneurs had desires which they cared to explore. By applying thoughtfulness and sensitivity, care bridged the relationship between the seemingly uneasy pairs of strategy and sensemaking. Although sensemaking for the most is said to be retrospective, it was the caring endeavors to realize the site, to unfold the innovation, which made it come into being. It was the caring approach that made them translate the strategy and enrol new resources, thus finding a sustainable business concept.

In the narrative, I described how the e-realtor entrepreneurs translated the business concept from being merely a web-based meeting place toward being a more traditional realtor, although with a generic internet-based infrastructure. In the initial phases, the desires of the entrepreneurs were fueled by the opportunities provided by the Internet. This was indeed a process of sensemaking where cues, in the form of technological opportunities, market disequilibria, were extracted and formed a basis for enactment and further institutionalization of the emerging network of resources. Both entrepreneurs had internalized knowledge within the ICT domain as a result of their formal education in law and the management of technology, although not having the technical competences related to ICT. Instead they were building a strategy around a concept utilizing what they found was disequilibria in the market, and outsourced the creation of the more tangible website.

The social capital in the form of "entrepreneurship-desires" seems to be important in order to form and sustain the network – and may be an important source for the caring approach. The entrepreneurs' confidence in their own ability to realize the business concept was important. But more important was their competence to "read" the limits of the Internet, thus translating the business concept toward a little "e." This meant that they did not have the same destiny as many of their kindred spirits within First Tuesday: becoming insolvent as a result of a unilateral concentration on the opportunities provided by the Internet.

The second theoretical perspective I wanted to explore was how collective sensemaking plays a role in processes of innovation. Hypothetically it was suggested that the likelihood of a network of actually succeeding in the creation of an innovation, is highly dependent on the network's ability to collectively produce sense in the process. RobinHus was started by the desires of two entrepreneurs, for whom the idea of creating "something on their own" in the business landscape was much more important than what this business should be about. That it was the realtor business they entered was by no means a planned strategy, but came into being as a result of a practical experience when one of the entrepreneurs engaged in his own estate transaction. The "pushy environment" enforced a cue, in the form of an experienced "injustice" or disequilibrium, which were then enacted collectively by the two entrepreneurs. They set out to conquer this disequilibrum and created a business. Almost like David against Goliath, it made sense to them and the employees they enrolled in the network in the consecutive years: They applied a brand new technology in order to compete against much bigger competitors in terms of size, turnover and resources. They did not use the narrative of David against Goliath, but they had considered themselves as the "impertinent boy in the class" in the early phase. This cultural identity was still an important character of the organizational culture in the office in Copenhagen as well as in the relation to the independent realtors. It was an important narrative to create a collective sense of identity. However, it also was a dilemma since it was not congruent with building a client-trust-relationship. On one hand, customers liked the notion of supporting somebody who was up against a large conservative business culture (traditional realtors), however these potential customers, on the other hand, were reluctant to engage with the impertinent boy.

While network and sensemaking together formed the fundamental perspectives in this chapter I find that the concept of "care" turned out to be an important conceptual tool. The concept of "care" is familiar to most of us, and we have a "feeling" that we know what care is about. However, care also implies a very broad range of meanings including – to mention a few – assiduousness, thoughtfulness, sensitivity, consideration but also anxiety, trouble and concern. It might be argued that this lack of conceptual accuracy makes it an odd concept in (micro)sociological studies. On the other hand, the connotations we attribute to the concept of care may be highly accurate and descriptive for the processes of innovation as well as entrepreneurship, where the balance between strategy/strategizing and sensemaking is important.

NOTES

1. The research conducted is part of a project that focuses on the innovation and role of ICT-based knowledge services. http://www.eservice-research.dk/. See also (Hagedorn-Rasmussen, 2006) and (Hagedorn-Rasmussen *et al.*, 2005).
2. Information and Communications Technology.
3. A more in-depth theoretical account is beyond the scope of this chapter. However, the approach here is inspired by the work of Czarniawska (1992/7) who also lends to the sociology of actor network theory.
4. A stream at the EURAM conference 2006 was devoted to exploring strategy as practice in which these issues were highlighted. Also see special issue of *Journal of Management Studies* 2003 (January).
5. The empirical research of an e-realtor, RobinHus, was conducted in 2004–5. It is based on interviews with each of the entrepreneurs, the customer relations manager, the IT developer and a real estate agent who is part of the network associated with RobinHus. Each interview lasted approximately 1½ hours. As part of the interview with the customer relationship manager we spent an hour together in virtual space.
6. This free service implied a "market place" on the site only. However, the concept – as described – came to include a broader range of services which were charged for. However, this charge was a flat rate as opposed to traditional realtor's commission. The RobinHus services implied a degree of self-service which made it possible to keep this charge low comparative to most commissioned charges.
7. The entrepreneurs have described this both in interviews but also in newspaper articles.

REFERENCES

Bijker, W.E. (1997), *Of Bicycles, Bakelites, and Bulbs – Toward a Theory of Sociotechnical Change*, Cambridge, MA: MIT Press.

Burrell, G. and G. Morgan (1979), *Sociological Paradigms and Organizational Analysis: Elements of the Sociology of Corporate Life*, Aldershot: Gower.

Czarniawska-Joerges, Barbara (1992), *Exploring Complex Organizations: A Cultural Perspective*, Newbury Park, CA: Sage Publications.

Czarniawska-Joerges, Barbara (1997), *Narrating the Organization: Dramas of Institutional Identity, New Practices of Inquiry*, Chicago, IL: University of Chicago Press.

Dougherty, D., L. Borelli, K. Munir and A. O'Sullivan (2000), "Systems of organizational sensemaking for sustained product innovation," *Journal of Engineering and Technology Management*, **17** (3–4), 321–55.

Hagedorn-Rasmussen, P. (2006), "Making sense of "E-knowation" – exploring the relationship between emerging strategy, innovation and entrepreneurial nets of critical capabilities and resources," working papers from Project E-service – Knowledge Services, Entrepreneurship, and the Consequences for Business, Customers and Citizens, vol. 8, Lyngby: Center for Information and Communication Technologies, Technical University of Denmark.

Hagedorn-Rasmussen, P., A. Henten and R.B. Møller (2005), "E-baseret ejendomsformidling i RobinHus," Working papers from Project E-service – Knowledge Services, Entrepreneurship, and the Consequences for Business, Customers and Citizens, vol. 1, Lyngby: Center for Information and Communication Technologies, Technical University of Denmark.

Hamel, G. and C. Prahalad (1994), *Competing for the Future*, Boston, MA: Harvard Business School Press.

Jennings, P.L., L. Perren and S. Carter (2005), "Alternative perspectives on entrepreneurship research," *Entrepreneurship, Theory and Practice*, **29** (2), 145–52.

Kirzner, I.M. (1992), *The Meaning of Market Process: Essays in the Development of Modern Austrian Economics*, London. Routledge.

Lengnick-Hall, M.L., and C.A. Lengnick-Hall (2003), "HR's role in building relationship networks," *The Academy of Management Executive*, **17** (4), 53–64.

Merton, R.K. and E. Barber (2004), *The Travels and Adventures of Serendipity: A Study in Sociological Semantics and the Sociology of Science*, Princeton, NJ: Princeton University Press.

Salaman, G., J. Storey and J. Billsberry (eds) (2005), *Strategic Human Resource Management – Theory and Practice*, London: Sage.

Schumpeter, J. (1975 [1942]), *Capitalism, Socialism and Democracy*, New York: Harper.

Snow, C. and R. Miles (1992), "Managing 21st century network organizations," *Organizational Dynamics*, **20** (3), 5–20.

Sundbo, J. (1998), *The Organization of Innovation in Services*, Copenhagen: Roskilde Universitetsforlag.

Sundbo, J. and F. Gallouj (2000), "Innovation as loosely coupled systems in services," in J.S. Metcalfe and I. Miles (eds), *Innovation Systems in the Service Economy*, London: Kluwer.

Weick, K. (1979), *The Social Psychology of Organizing*, Reading, MA: Addison-Wesley Publishing.

Weick, K. (1995), *Sensemaking in Organizations*, Thousand Oaks, CA: Sage.

Wright, P.M., B.B. Dunford and S. Snell (2005), "Human resources and the resource based view of the firm," in G. Salaman, J. Storey and J. Billsberry (eds), *Strategic Human Resource Management: Theory and Practice*, London: Sage.

13. Making innovation durable

Connie Svabo

This chapter examines innovation in the context of the small scale fashion industry. The chapter tells a story of the struggles of translating innovative ideas into material forms, and shows that the work of innovation consists of continuous attempts to create material order. Hardy negotiations are involved when innovative ideas are given durable form. The process of materialization is a complex process, where abstract ideas may meet material obstacles and textural boundaries. Innovation, thus, is a continuous process of negotiating material realities.

The main points of the chapter are that innovation is a continuous process of working for order. Very often, material objects are central in fixing and materializing an innovative idea. This chapter argues that it does not make sense to talk about innovation without focusing on the materialities which the innovative efforts are centered around. Innovation is viewed as a process which works against a background of fluidity, of continuous decay and disintegration. Central to innovative efforts are two things: materializations – working at and with physical objects – and conversational processes which aim at sharing perspectives on the innovation activities. Processes of innovation are processes of social and material conversation and negotiation.

The concept of innovation with care points to relations between the diverse sets of actors which may participate in innovation, and the advice that these relations must be handled with care – that they must be worked at and bridged together. This chapter suggests that this bridging is undertaken with explicit attention to the role of material entities. Material objects are viewed as potential stabilizers, they are viewed as the boundary objects which may give concrete form to innovative ideas, but they are also viewed as entities which are fragile. Their cohesion may be obstructed, and their materialization may fail to emerge, and even once they are constructed, they may easily fall apart again. This shows that innovation is a continuous process, and that maintaining an innovation is just as crucial as the initial birth of an idea. Innovation is portrayed as taking place in tightly knit, inseparable, and co-creative socio-material relations in continuous making and decay.

This chapter wants to demonstrate the role and importance of concreteness, of form, when working with innovation and creation. Material artifacts are central "stabilizers,"[1] they may create durability, and they may be the site of negotiation between diverse organizational realities.

The theoretical vocabulary which is used in the presentation of this case study is, to a large degree, inspired by actor network theory (the sociology of translation), which briefly will be introduced. Furthermore, along the way, the case study is related to three modes of innovation: entrepreneurial innovation, open innovation and strategic reflexive innovation. In this text, each of these modes are used as a narrative figure which helps reveal distinct aspects and interpretations of the innovative case. For an introduction to these figures and their background in innovation theory, see Fuglsang (Chapter 5 in this volume).

ACTOR NETWORK THEORY

Actor network theory is an intellectual tradition, which has its base in "a ruthless application of semiotics," as John Law puts it (Law and Hassard, 1999). Latour, who, along with John Law, may be counted as one of the VIPs in this intellectual tradition, stresses that actor network theory (ANT) is a method which may be used to study all sorts of things. This method emphasizes that when looking at social phenomena, for instance innovation, we should look at the oscillation between fluidity and durability.

Latour's picture of the social world is that it is "fluid," it is constantly "decaying," shifting and changing (Latour, 2005). Latour stresses that sociology should seek to understand movements between fluidity and durability, the central questions being: how do stabilizations emerge? How does order come to be? And, in this context, how is innovation made durable?

Latour points at two phenomena which make constantly shifting, fluid interactions more durable (Latour, 2005: 68). These are "things" and "social skill." Of these two, it is especially the first that has been the center of attention in work done within an ANT tradition, probably because this approach is quite radical in its claim that objects too have agency. Latour says: "It is always things – and now I mean this last word literally – which, in practice, lend their 'steely' quality to the hapless 'society'" (Latour, 2005: 68).

Latour reveals some tricks to make these otherwise mute things talk. Looking at innovations is one of the tricks. Looking at innovations, distances, accidents, breakdowns, strikes, history and using fiction may be useful in analysing and in writing the stories of the heterogeneous relations from which social and material reality is woven. Focusing on innovations may bring "the solid objects of today into the fluid states where their connections with humans may make sense" (Latour, 2005: 80 ff).

Further advice given as regards method to be employed for analysis is quite simply put: "Just describe the state of affairs at hand" (Latour, 2004). This is attempted in the following – using vocabulary which has an ANT-feel to it.

INNOVATION AS A MATERIALIZATION OF IDEAS

This is the story of an innovative project, "Sidecar," which was both a success and a failure. The project was an innovation in the sense that it was an economic and social activity, the purpose of which was to add value to the firm and to the participants in the project. The value was to be generated by bringing together new designers and an established production and distribution network. The project was a success since the primary goal of producing a marketable collection of designer clothes and presenting it to buyers at a fashion fair was achieved. It was a failure because internal difficulties between the project manager and the lead designer led to the project manager quitting her job, and this again led to the owner closing down the project.

The firm, which no longer exists, will in this chapter be given the fictional name of IDT. Sidecar IDT consisted of four people, the two most important in this context being the owner and the lead designer. The owner was a classical, patriarchal, "old-school" entrepreneur who still balanced the firm's accounts using paper and pen. He was in his late 50s, had extensive business experience, and had been self-employed for most of his life. He very carefully monitored income and outcome, and, metaphorically speaking, did not pay bills, before income had been generated. He had several years of experience in the fashion industry, among other things having owned a retail store. He was the owner of IDT, and had been cooperating with the lead designer for five years before IDT was established.

The lead designer was in his early 40s. He was self-taught, and had worked with fashion design for more than 20 years. He was a name in Danish fashion design due to a brand he launched in the 1990s.

The IDT business relationship started by chance, the lead designer was renting store-space in a building which belonged to the owner. The lead designer had economic problems and the owner helped him out by buying his firm and his new brand. Together they built a small, but successful and growing international firm, which sold designer clothes.

The background for Sidecar was the owner's retirement plans and the entry of the project manager, a self-employed woman in her late 20s with a business degree, broad experience with project management and an interest in the fashion industry. She was a remote acquaintance of the owner, and her role in the firm was to do the books in cooperation with the owner, and thus gradually take over the financial management of the firm. Parallel with this, she was to be responsible for a development project to which the owner had gotten the idea, namely Sidecar.

The idea was to cooperate with a handful of young fashion designers to show their clothes at a fashion fair, and hopefully sell their models to international distributors. This idea was conceived because one of the primary difficulties for new designers in Denmark is to get access to production – and distribution networks. As a relatively well established, internationally selling firm, IDT had both.

ENTREPRENEURIAL INNOVATION

The metaphorical figure of entrepreneurial innovation captures important aspects of the Sidecar project. It is the story of the dynamic person who seeks to create something new, based on his experience with fashion industry, and his perception of structural hindrances in this sector. The entrepreneur in Sidecar was the owner and everyday manager of the firm, a business man on his way to retirement for whom IDT was a form of "leisure business."

His motivation for launching the innovative project was a combination of profits, the game of developing a firm, and the possibility of helping young designers' entry into the business. As such, and in accordance with entrepreneurship theory, his motivation was a combination of economic and social value. This figure of entrepreneurial innovation tells us a classical story of entrepreneurship, but it also paints a quite incomplete picture, where other actors and their motivations are toned down.

ASPIRING FASHION DESIGNERS

When a fashion designer attempts to "go commercial" there are several things that have to work out. First, the designer has to produce a "collection" consisting of a number of "styles." The collection is a whole range of articles: pants, skirts, blouses, shirts, jackets, knitwear, and perhaps accessories. A collection may also be a range of garments, which are made with the same technique, for instance, seven different styles of knitted shirts and skirts. As design, the quality of a collection may be judged for its "form language" for its coherence and contrasts, for its themes, and for its ability to give an appealing aesthetic impression.

This is all about design, and it involves a lot of imaginative work (Strati, 1999), for instance, the work of making a collection look convincing. This may be done using diverse media. One of the most important is doing samples. The designer manually produces, often on her own sewing or knitting machine, samples of the clothes. These samples may be demonstrated on models, at shows for example, and in pictures. This is where stylists and photographers enter. Again with imagination and creativity – making things look, appear, perform.

Suchman writes that building bridges may be viewed as building stable artifacts, and that this (building stable artifacts) "involves the accomplishment of alignments across heterogeneous human and nonhuman elements," the work of bridge-building may be viewed as "persuasive performances that both rely upon and reflexively constitute the elements to be aligned" (Suchman, 2003).

The same can be said about fashion design. This concept of persuasive performances grasps what a lot of the pre-production work of small-scale fashion design is about. It is about creating persuasive performances with very small budgets.

There are several fixed points for these persuasive performances. One of them – and a very important one at that – is the fashion fair. In fashion design in Denmark, an important fair is the one which takes place in Øksnehallen, twice a year. Throughout the last decade, Denmark increasingly has become a fashion city, and actually three fairs take place at the same time. CPH VISION in Øksnehallen is for "designer" clothes. At Bella Centret the fair is more oriented toward large scale mass production and consumption. And right next to Øksnehallen, an old factory building houses an "underground" designers come-together: The Fanny and Foxy Fashion Fair. Fanny is a woman, Foxy is a dog.

These are the fairs, which are so important because this is where buyers shop around, look at collections and styles, and place orders. This is where the purchaser from the large Japanese chains of department stores may see the designers' items, or where the purchaser from Harvey Nichols may change a designer's destiny by placing an order. The air is loaded with hopes and money, and there is an intense atmosphere of show and tell, and of power and judgment.

Everybody hopes for orders. Large orders means large scale production, which again means that cost per unit goes down, and revenue goes up. If an order is placed by Moshimoshi or Harvey Nichols, because there are so many stores, the demand is usually large enough for industrial production. With a large order, the designer has no problem convincing the production network to give high priority to their order. This enhances the possibility of delivering on time, and thus reduces the risk of complaints.

But the people browsing around the cubicles in Øksnehallen are not all from Moshimoshi or Harvey Nichols. There are also the much smaller, but nevertheless important, Nordic shop owners. People who own one or two shops, and who are out to gather things to put on their shelves and hangers. Things that will go well together, and which they expect to be able to sell. An order from one of these is nice for the hopeful, young fashion designer (and also for the more established designer) but these orders are never very large, and they therefore may leave the designer with a problem of production. If the designer gets a small order for a garment, and this order is not followed up by other orders for the same garment, the designer faces a production problem. The styles that are displayed at the fair are primarily industrially produced prototypes, and placing a production order for 20 pieces will very often be rejected. This leaves the designer with a huge problem. An order was made, but how will she deliver?

This is a rope which entrepreneurial fashion designers balance on: should they get their prototypes industrially produced, and thus be sure that what they are able to deliver is identical to what is on display, and thus also have the ability to deliver large numbers of each garment? Or should she make the prototype herself, assume the huge work load of production, and thus risk late deliveries and production difficulties?

MATERIALIZING THE IDEA OF A SWEATER

There is a long process before a garment gets as far as a fashion fair. This process will be exemplified by telling the tale of a sweater which was made for the Sidecar collection.

The sweater started out as an idea, a sketch on a piece of paper. A student at one of the Danish design schools had gotten an idea for a sweater. She had drawn this sweater on pieces of paper and in various forms, and the shape was promising. Its form fitted conceptually with other forms that she was experimenting with, and she chose to use it as one of the styles that she would present at her graduation. In order to do this, the idea of the sweater had to be translated from the two-dimensional form of a drawing to the three dimensional form of an actual sweater. The designer did not knit, but she knew an elderly woman who did, and this woman was willing to help her out. She hand knitted a prototype of the sweater. The designer was content, and was able to use it as a style in her graduation collection. It made a persuasive performance, and her work was accredited.

Enter Sidecar.

In a recruitment process aimed at getting people to participate in a development project, which was giving young designers the possibility of trying to make commercial clothes production with their own brand, and under the Sidecar umbrella, the design prototype, among other styles, made a persuasive performance again. The project manager and the renowned lead designer of the firm IDT found the sweater design very promising. The young designer, along with five other female designers, was offered participation in the development project, and she accepted.

SELECTIONS

The aim of creating the Sidecar collection was to make a selection of items which would be presented at the February fashion fair. The process of selecting the items was carried out with help from external sources. The six designers were asked to present their styles to a trio consisting of two important actors on the Danish design scene and one customer representative. These actors participated *con amore* in a two hour test-case, where each designer presented their collection, and received comments on their collection as a whole, and on the individual styles. The name and network of the lead designer was central in getting these important people to participate.

This qualifying round had several positive effects. It served two purposes, it helped spread the word about this new and interesting, innovative (!), project in the design community. It "outsourced" the selection process, securing an external, design-based judgment of the designers' contributions and reducing the personal, intraorganizational discussions on this delicate issue, imitating the manner of judgment which the young designers were used to from school, and making the criteria of judgment ones of "good design," and thus also reducing the potential, personal conflicts related to a process where some styles were deemed "in" and others were deemed "out."

OPEN INNOVATION

This qualifying session is a useful example of open innovation planning (Chesbrough, 2003) which has taken external sources into consideration and has involved these actively in the innovative process. This was done by organizing interaction with a person who was perceived as being representative of "a core customer" and by involving a professional network which was perceived as being representatives of "knowledge about design trends." In accordance with the literature on open innovation, this "qualifying round" dispersed the power of one professional (the lead designer) and delegated it to a broader net. This made the process of selection an open and explicit one, where "the rules of the game" were verbalized and written down. This was done with the primary purpose of communicating with the external participants, but also had positive effects on the designers' understanding of what was going on. What took place may be viewed as a decoupling of professional experience from decision-making, but a more appropriate description would be to talk of a delegation involving a broader range of professional and customer perspectives. This provided input into the decision-making process, and gave the lead designer and the project manager a common point of reference in the process of chosing styles, and in the process of communicating these choices to the designers.

It was a "rule" that all of the designers would be able to participate with a garment.

As demonstrated above, there are several organizational strengths to such a process. There is also an important weakness. The selection process led to the fragmentation of some of the designers' collections. Some designers had several styles chosen, and thus their design identity, the coherent expression of their styles was intact, but for others, who only had one or two styles selected, their garment was no longer part of the coherent expression which they were attempting to construct, but rather, it was "just a piece of clothing" among others. The overall picture was that of Sidecar, which in its essence was a conglomerate of different design expressions, and thus not a coherent and singular design statement. Retrospectively, this is one of the aspects which the project manager would consider changing if the project was to be carried out again. It may have been a wiser choice not to "bet on all of the horses" but to have chosen one or two designers and their entire design collections.

The figure of open innovation helps portray Sidecar as an attempt at innovation, which involved interactions with many and changing actors, not so much in getting the idea, but most definitely in carrying out the project. A range of actors were involved. First and foremost, the six designers who were chosen to participate in the project, but also a range of other actors. The design schools who helped point out which designers could be interesting to recruit, the nephew to the entrepreneur-owner-manager who is a professional photographer and who owed a favour and thus could deal with a low (no-)budget production of marketing material. The husband exhibit designer who helped set up the fair booth and make ends meet, and others. These actors were not *the* source of innovation, but telling the story of the innovative project, its successes and failures, without these actors does not make sense. They were crucial contributors both in relation to sources of creativity and in carrying out the innovation. External sources, in the Sidecar project, were not only aspects which would need to be taken into consideration in innovation planning, they were central drivers. Innovation planning could not have been carried out without these external sources.

MATERIALIZING A SWEATER (PART 2)

Now a new round of translation[2] started. The first round was the translation of a sketch into a knitwear to be displayed for graduation. This process had been relatively easily monitored. The designer had purchased yarn in a random store, and had given it to the woman who knitted the garment. The designer had plenty of time to follow the production process, and to instruct for small adjustments. When the knitwork was finished, the designer herself mounted a zipper, which she also had purchased in a random store in the city where she lived.

It turned out that things were much more complicated in getting the second prototype ready. This prototype had to be an industrial prototype. A prototype that gave a true picture of the garment that a shopowner would receive if an order was placed at the fashion fair. And a prototype which also was the basis for calculating production costs, and thus retail sales prices.

The helpful lady who had made the school prototype was not part of a large scale production network, so she could not do it. Things were a bit more complicated now. The designer had gotten a full time job, she would like the sweater to get to the fair, but her resources for finding samples, looking for production possibilites and monitoring the production were very limited.

The project manager at IDT nevertheless, was quite intent on getting this sweater to the fair. She liked the sweater and believed in its qualities and its market value. So she took on the work of getting a prototype made. She was responsible for all of the prototypes in the Sidecar collection, but some of the designers were quite involved in these processes, and not all of the prototypes were as demanding as this sweater.

Project manager's first difficulty was to find somebody who could produce knitwear industrially.

"Getting prototypes made from large international production firms is very difficult," said the lead designer, who happened to be going to Italy where his own garments were produced, but "this firm will not be able to do knitwear," said the lead designer. "They do know someone who could do it, but relations are a bit tense, our payments to them have not been stable, and we are not in a position to ask for favors." According to the designer, the project manager was best off if she attempted to find production facilities in Denmark. "Knitwear is a niche form of production, and getting production done in Denmark is not very easy either," said the lead designer, and "No, unfortunately I don't know of anyone who can do the job."

The project manager rang around and ended up getting hold of a teacher at a design school who provided her with a name of a man who had industrial knitting machines, and who did good work, said the teacher, and who to her knowledge was willing to do stuff for aspiring designers.

The project manager called the man, introduced herself and the firm she was calling from. "Oh, yes, yes I know the lead designer," said the man on the phone, and the project manager was baffled, and the man continued, "I did a lot of stuff for him some years ago."

The knitmaster was willing to do a prototype. This was a very important milestone. The machinery was there and the man was there. Now a comprehensive job of choosing yarn, choosing zippers and making knitting instructions was started. These were translation processes. Creating knitting instructions for a machine knit was a difficult work of translation for the designer. She was being asked to speak knitting machine language, a language she did not know. The lead designer helped out in this process. Furthermore the designer was to choose yarn, for the yarn she had purchased in the random store for the school prototype was not the kind of yarn that the knitting machines used. The knitmaster had some specific firms with which he cooperated, and the designer was asked to choose some of these yarns. Samples were sent back and forth, and the same process was carried out for zippers. Everything was carried out operating with tight deadlines, and with a designer who was occupied full-time sketching shirts for a large-scale clothing production firm. The prototypes turned out differently from the way the designer wanted them, the sleeves were placed too low, and the neck was too long. She was not happy, but alright then: the prototypes were there, and the project manager was content, they had something to present at the fair.

The struggle of translating a sweater prototype which had made persuasive performances into a new sweater prototype, which had to make other persuasive performances was a struggle of material form. The process of materializing a knitted sweater is a process which involves entities of heterogeneous character. Yarn made of pure wool behaves differently from a thinner mix of wool, acrylic and polyester. Zipper factories do not always have all zippers in all colors available for delivery within short time periods. And professional knitting machine language is very unlike the language of a woman doing a singular prototype for a school show.

The process of materializing the idea of a sweater was a translation process which met several obstacles. One important obstacle was getting access to machines, there were actions which may be interpreted as obstruction by a designer who was not willing to share his knowledge about machines, about production facilities and competences, with other designers. But also very basic and practical things such as getting the right wool and the right zippers to be at the right place at the right time.

In fashion business, hindrances to materialization are "deadly," be they due to obstructive colleagues, unreliable producers, shortages or other things. A very large part of all activity within fashion business is about materializing ideas. About transforming and translating abstract ideas into concrete styles, which because of their material qualities, texture, style and form convince a buyer.

And what happens in that crucial moment where this materialized idea meets a hand, when it is fondled, touched, held out as it displays itself on a hanger at a fashion fair?

Using Latour's terminology one can say that the designer hopes to delegate an action program to the non-human, the thing, the sweater. But not only into the sweater, for that is a weak constellation. Going to a fashion fair with only a sweater, laying it out on the floor, a single item, that won't do. A series of other entities are aligned to make a durable and performing net: hangers for one.

Hangers are very, very important. They have to be neutral, signal the right style, and have an appropriate weight. They have to be able to hold the garment, and with some garments for females, this is quite a difficult task. This is where another important entity comes into play: rubber bands. Regular, little rubber bands may be wrapped around the hanger, providing friction and a little bump on the surface of the hanger, making the garment stay in place.

But not only these entities are important, so is the space around the sweater.

ALIGNING SPACE

At a fashion fair the material arrangements tell a great deal about the size and economy of the firm. The large and established brands, with extensive experience from fashion fairs, have carpenters, they have people who move in and set up everything before the sales team enters to do their professional work. In the small scale firm, the independent "low-budget" designer must take care of everything from a to z. It is a strenuous business, this business of arranging a performative fashion fair stand. It is strenuous in the sense that the garments arrive to a fluid reality, a reality which is not in an appropriately stabilized form. Not very much is fixed in advance. There are a lot of artifacts that have to be arranged, a lot of stabilizations to be made. And this is demanding, especially of the novice. These processes can be viewed as "fights for durability," struggles to create stable, orderly and performative materialities.

An extensive list can be made of actions that were carried out in the intense 24 hours which were at disposal for setting up the stand at the fair.

In the Sidecar project, there was an intensive negotiation about space. How much space and which space. The firm's Fashion Fair stand all in all was 9 m². Sidecar was given 3 m², the lead designer had 6 m². This was decided by the owner and lead designer while the project manager was on vacation. This issue and other issues were questioned by the project manager at the fair. Have you ordered cables and lights for us as well? Where do we get cables from? May we move you back just a little?

Three square meters is not very much for six designers and a project manager.

The Sidecar Fair Stand was the ultimate materialization of the innovative project. This was the arena, where the tangible artifacts which the designers and project manager had worked at for an intense period of six months, were to perform.

And they did, perform. Orders were placed for sweaters and knitted skirts. Not very large orders, but orders. The Sidecar collection was made, and purchasers were touching it. It was there, on two rows of hangers. A rich array of material entities had been constructed, gathered, and produced in order to do a persuasive performance in this arena. Twenty-four prototypes had been designed, arrangements had been made, drawings had been flown to Italy, Portugal and various sites in Denmark, garments were produced, garments were shipped, tags were designed, washing instructions were noted down, color availability was cross-referenced, material delivery times considered, prices calculated, style sheets layouted, samples obtained and fragments glued into display booklets.

SIDECAR POSITIONING

The Sidecar project and the work of creating a collection for the Fashion Fair illustrates one of the aspects of innovation with care, namely the aspect of *positioning*, meaning that companies must create relationships to multiple and distinct actors with different starting points, approaches to and interests in innovation.

One of the strengths in the project was the obvious interest the two main parties had in each other: the designers were strong on design but lacked business knowledge and access to distribution and production networks. The company had business knowledge and distribution and production networks, but was interested in developing a new design range. These backgrounds and stories positioned the parties in relation to each other, and these positions and the relations between them were active drivers in the innovation process. The distinct concerns and interests of the involved parties were voiced in collective meetings, in person to person conversations, through written documents, through contractual agreements, through time schedules and production plans, which were used as graphic depictions which signalled and asked for commitment and clarity in work processes.

The mutual stories and the material form they are given are central. In Sidecar, they were written on the webpage, but most importantly they were written in newspaper articles, in fashion magazines and fashion industry magazines. These materialized stories which told of the project idea and philosophy, which mentioned the names of the participants and of the future plans for the project became tacit and tangible signs that the project was real. These stories verbalized what the project was about, and who participated in it. This type of communication had external goals but most certainly also internal effects.

Telling collective stories, and giving these concrete and material forms helps build relations, and helps engage people into continuously stabilizing the innovation and making it durable. The platforms for collaboration are built into the objects, and the mutual stories are created by listening, talking and writing.

INVOLVEMENT IN SIDECAR

Another aspect of innovation with care – that of *involvement* – was also central in creating the Sidecar collection. This was made explicit in the initial conversations and sharing of perspectives and interest. The project was on a very low budget, and in many ways worked in the same mode, as the designers would have done without the firm – creating professional looking performances with low-budget means and making things possible through networks. Important devices for maintaining involvement were communication, giving information on new developments, hosting meetings and delegating tasks. The designer involvement was not difficult to gain because of their obvious personal interest in the project, but involvement was very difficult to gain from the lead designer, who seemed to think that he had a personal interest in *not* helping the project development. How his involvement could have been created is still an open question.

TURNING INNOVATIONS INTO PREMISES FOR PEOPLE'S OWN ACTIONS

Fuglsang and Sundbo (2005) mention that how people can turn innovation and innovation systems into premises for their own actions must be taken into account, when the right environment for innovation is engineered.

In which ways was Sidecar a story of people turning innovation and innovation systems into premises for their own actions? In the relationships which were developed between the firm and the six designers. An agreement was made, and even though there was no exchange of money – this was all rendered into the future – social and material relations were established in a way where Sidecar became a very important premise for the actions of the designers. Sidecar was the "institutional" context which they referred to, Sidecar was their possibility of getting to a fashion fair, they worked at creating collections with the purpose of presenting them in this innovative project, and the premises of the firm and the project became important references in their activities.

Using actor network theory one can say that a boundary object emerged (Star, 1989). A boundary object is an object which simultaneously belongs to distinct social worlds (distinct social practices or distinct actor-networks). Conceptually, boundary objects have an interesting double-ness – they are flexible in the sense that they are able to belong to these distinct worlds, and at the same time they have durable qualities in the sense that parties from distinct social worlds seem to agree on at least some aspects of the boundary object. The emerging Sidecar collection acted as a boundary object between distinct communities. The established community of the firm was loosely coupled to other communities: To more or less established designers, some of whom worked from home, some of whom were members of communities with likeminded designer-aspirants with shared design- and small-scale production spaces. A relation was established to the design schools who helped point out potential participants, and a relation was established to the helpful participants in the "selection seminar," and so on.

The answer provided in this account of how people can turn innovations into premises for their own actions goes along the lines of stressing the role of "objects." This perspective shows how collective and social processes of making sense emerge in relation to a concrete and/or conceptual object. Involvement and participation emerge around this object, the object is central as a focus for collective and individual activity. The material world in this sense may be seen as something which mediates human activity. Boundary objects, such as the emerging Sidecar Collection, may traverse communities of practice. As such material objects may be viewed as artifacts that can help negotiate knowledge across distinct practical and professional domains (Carlile, 2006).

Not all objects may be boundary objects. Carlile in a collection of 65 observations of the use of different boundary objects has identified three characteristics:

> First, a boundary object establishes a shared language for individuals to repre-sent their knowledge . . . Second, an effective boundary object provides a con-crete means for individuals to specify and learn about their differences and dependencies – what is new – across a given boundary . . . Third, an effective boundary object facilitates a process where individuals can transform the knowl-edge being used. (Carlile, 2006: 115)

Boundary objects may be perceived as material objects which support actors to collaborate across boundaries.

INNOVATION AS PERSISTENT ACTION

The answer provided by this chapter as to how the proper "environment" for innovation can be constructed is through awareness and work on two levels: by materializations and through conversational processes which aim at sharing perspectives on the activity at hand – on co-creating social rep-resentations (Fuglsang and Olsen, 2004: 423). The figure of strategic reflexivity helps explore these issues further, but first a comment.

Judging from the literature, it appears that an important issue to consider is a question of causality. Implicitly it seems as if it is of special interest to trace the idea, where did the innovative idea come from, what are its sources, are they internal or external? Regarding strategic reflexivity, Fuglsang argues in the introduction to this book that strategic reflexivity also takes its starting point in external ideas and expectations to change, but this does not seem to apply to the Sidecar project. The idea, the source, was the patriarchal entrepreneur. But it seems a pity not to use the focus that is ascribed to external actors in developing a further understanding of the process of innovation. The thoughts coming from the notions of open innovation and strategic reflexivity can be used to clarify and stress the importance of involving external actors, in this case not as a source to the innovative idea, but as sources to the innovative process as a whole, in the everyday practical and managerial work of taking an idea and trans-lating it into marketable forms (for instance). In IDT/Sidecar, the idea of the innovative project was not generated by "expectations of change," but external sources were absolutely crucial in carrying it through. This focus on "the idea" which I attribute to both the figure of open innovation and strategic reflexivity can be criticized for a not uncommon tendency within social/organizational science to focus on the idea, the abstract thought, as initial and causal factor for social processes (Gagliardi, 1996). Tracing causalities in this manner is criticized by Latour, among others, for not grasping the complexity of actants which participate in social processes (Latour, 2005). My point is that maybe telling the story of innovation should not focus so much on the source of innovation (singular form), but more on the range of both internal and external actors and activities (multiple forms), situated in both temporal and spatial contexts, which cojointly make an innovation. Having made this point I will proceed with the tale of Sidecar as it is demonstrated through the figure of strategic innovation. The focus is on creating a platform for mutual learning and socialization.

STRATEGIC INNOVATION

The external inputs in Sidecar are not, as in Sundbo and Fuglsang's conceptual framework, macro-inputs. But the term external inputs is nevertheless useful, also on a micro-level. It can help illustrate some of the interactions which on a micro-level could lead to organizational learning. In this perspective, learning to a large extent is conflict-based (in a psychological sense). Learning is something which may be triggered by disturbances, conflicts and differences (Blackler, 1995; Blackler *et al.*, 2003; Larsen and Svabo, 2002; Raelin, 2000).

External inputs on an everyday level had the form of the project manager, who entered and by mere presence disrupted organizational space, rhythm and relations. The firm had previously consisted of four people, two of whom had managerial activities (the lead designer and the entrepreneuer-owner-bookkeeper), one shopkeeper/print assistant and one print/production responsible. The Sidecar project manager/bookkeeper-in-training came to be the third person on the managerial side. Spatially the firm consisted of two separate entities, the office and shop being one of them, and the small-scale print/production facilities being the other. These two spaces had separate entrances, situated 25 m from each other and opening out onto the same street. These two settings were close enough to each other to be "connected space," but at the same time the spaces were quite distinct. They were governed by different modes of action, and the people were distributed unevenly in them in the sense that the same person definitely belonged more to one area than the other. The space in focus here is the office and shop, which primarily was the space of the lead designer and the owner, but which with the entry of the project manager, also became her space. The bodily presence of another person in an established space created disturbances and

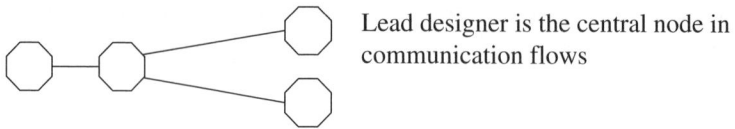

Lead designer is the central node in communication flows

Figure 13.1   Communication (a)

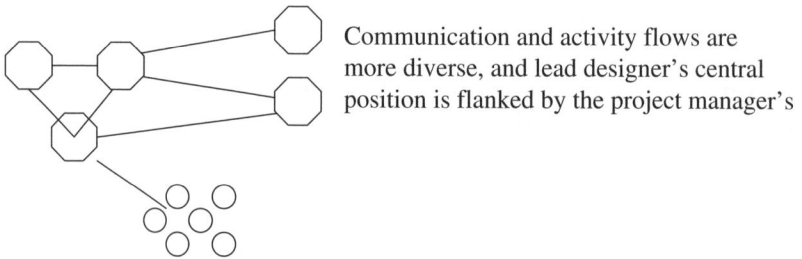

Communication and activity flows are more diverse, and lead designer's central position is flanked by the project manager's

Figure 13.2   Communication (b)

disruptions, also by the mere fact that there was space for two tables in the office – the owner's table and another desk, which was a more collective space, which could be used by the three other people, but which after the entrance of the project manager, became her space.

The presence of the project manager created disturbances in the rhythm and relations of the everyday life. A communicative arrangement, which previously had looked like that shown in Figure 13.1 now came to look more like that shown in Figure 13.2.

A great many things may be analysed from this depictation, the aspect which will be stressed here is the evident effect that the introduction of a number of external actors may have on power relations. External input in the form of a project manager, who is introduced as the person who in the long term is going to take over the management of the firm (when the owner retires) and who in the short term is project manager of a development project, does something to internal habits; affects them, and influences them. This is one of the aspects mentioned by Sundbo and Fuglsang (Sundbo and Fuglsang, 2002; Fuglsang and Sundbo, 2005) that strategic reflexivity takes into consideration external inputs and problematizes internal habits. On the other hand, internal routines are maintained and external inputs translated into something, which is perceived as useful by the experienced, professional culture. This did not really happen in Sidecar. The kind of socialization and learning that can be the positive outcome of disturbances did not emerge.

After the fashion fair, the project manager quit her job. She stopped struggling to create and maintain these innovative material arrangements. With that decision, the innovative project was called off. Sidecar crashed. The designers were given the opportunity of taking over the responsibilities for the orders that had been placed, and if they did not wish to do this, the orders were cancelled.

Mutual Understanding

Connecting this story of an innovation which crashed to the conceptual frame of innovation with care, and explicitly the stress attributed to mutual understanding in this frame, leads to the following speculations.

The concept of innovation with care suggests that the broader is the difference in backgrounds, the more difficult it is for people to understand and recognize each other's backgrounds, to balance goals and means, strategy and sensemaking. This makes differences in background an important explanatory factor – a factor which may lead to innovative failure or success. Whether this category of "professional background" is the most appropriate term to use, when attempting to capture the differences between people must be an empirical question (other relevant categories could be gender, age, and so on). Nevertheless, people have distinct and sometimes even contradictory impulses, emotions, feelings, intentions and meanings. And the point made by this chapter is that working with or against these backgrounds, people negotiate themselves "to order" in materiality, through material forms.

Mutual understanding in some respects was central to the success of Sidecar, and lack of mutual understanding in some respects was central to the failure of Sidecar. Conversational work – having meetings, talking about visions, getting stories printed in newspapers, and so on – was very important in building the relationship between the young fashion designers and IDT. And equally, the difficulties in building a cooperative relationship between the project manager and the lead designer were important in the failure of Sidecar. Evidently, thus, this chapter does not deny the role and importance of mutual understanding. But what it really wants to do is address primacy to materiality and to demonstrate how – with or without mutual understanding – working with innovation is a continuous process of negotiating and stabilizing material realities. The point made by this chapter is that innovation can be made without mutual understanding; the Sidecar Collection did get as far as the show, but it cannot be made without materializing the innovative idea.

MATERIALITY FIXATES INNOVATION

Materiality mediates human activity, and in these processes of materiality mediating human activity, human knowing and being are also being shaped (Pratt and Rafaeli, 2006; Pels *et al.*, 2002).

This chapter stresses the material and tacit forms as crucial fix points. These objects are loaded with energy, they are the objects which designers, project managers and knitmasters spend hours toiling over, they are the objects which fill working life activities, and they may even be the objects which decide future destinies.

Fixation of material form is thus also a fixation of identities and of power. In an organizational reality negotiations about material forms – for instance negotiations about the distribution of square centimeters in a fair booth – are negotiations that may reveal organizational identities, and distinct conceptions of reality. The thing, the material form becomes the place, where multiple organizational identities are in negotiation or at war (Pratt and Rafaeli, 2003).

An innovation may be seen as an artifact which is continuously engineered and maintained in a stable form. Attention in processes of innovation should thus be given to working with both social and material relations, which are hoping to produce effects, which may be perceived as innovative.

This case study hopes to show how the struggles to create and innovate are continuous struggles of making durable material arrangements. The work of innovation is work at material form, of people working at distinct and similar artifactual objects, and it is a very materially confined and textural work of creating performative shapes. This chapter points toward innovation as a process which must continuously be actively sustained – materially and relationally. It points at innovation as a process, a struggle for orderly material arrangements.

This does not deny the value and importance of communication, of social skills, of listening to the coworker, recognizing distinct backgrounds and so on. These communicative, conversational activities are definitely important, and they may contribute positively to the stabilization of material form, but they are not *the* explanation of how innovation is achieved. Innovations can be achieved without mutual understanding, and mutual understanding does not necessarily lead to innovation.

This case study hopes to contribute to the overall concept of "innovation with care" with this story of how – in practice – diverse actors struggle with creating durable forms, and most specifically how the work with these forms are continuous and emergent processes of negotiation and conflict. This view shows innovation as tightly knit, inseparable and co-creative socio-material relations in continuous making and decay. Processes of

innovation are processes of social and material conversation and negotiation. Differences and similarities in understanding are weaved – or perhaps
knitted – together in more or less stable materialities.

NOTES

1. Stabilization is a central concept, and regards how durability may be achieved in a social
   group by the creation of common knowledge and understanding as regards for instance
   the innovation.
2. Translation is a central concept in ANT, which is also the reason that this intellectual tradition is sometimes called the sociology of translation. Processes of translation are about
   changes in network relations which may change the relations between actors, may activate
   new actors and change the entity which is the focus of the network. The concept both
   means translation – literally and – displacement. The extent and direction of translations
   depend on the resistance provided by other actors in the network. A translative process
   may lease to a new constellation of actors.

REFERENCES

Blackler, F. (1995), "Knowledge, knowledge work and organizations: an overview
and interpretation," *The Journal of Organisation Studies*, **16** (6), 1021–46.

Blackler, F., N. Crump and S. McDonald (2003), "Organizing processes in complex
activity networks," in D. Nicolini, S. Gherardi and D. Yanow (eds), *Knowing in
Organizations: A Practice-based Approach*, New York: M.E. Sharpe.

Carlile, P. (2006), "Artifacts and knowledge negotiation across domains," in
A. Rafaeli and M.G. Pratt (eds), *Artefacts and Organizations: Beyond Mere
Symbolism*, London: Lawrence Erlbaum.

Chesbrough, H.W. (2003), *Open Innovation: the New Imperative for Creating and
Profiting from Technology*, Boston, MA: Harvard Business School Press.

Dugdale, Anni (1999), "Materiality; juggling sameness and difference," in J. Law
and J. Hassard (eds), *Actor Network Theory and After*, Oxford: Blackwell.

Fuglsang, L. and P.B. Olsen (2004), *Videnskabsteori i samfundsvidenskaberne. På
tværs af fagkulturer og paradigmer*, Frederiksberg: Roskilde Universitetsforlag.

Fuglsang, L. and J. Sundbo (2005), "The organizational innovation system: three
modes," *Journal of Change Management*, **5** (3), 329–44.

Gagliardi, P. (1996), "Exploring the aesthetic side of organizational life," in
S.R. Clegg, C. Hardy and W.R. Nord (eds), *Handbook of Organization Studies*,
London: Sage.

Gherardi, S. and D. Nicolini (2003), "To transfer is to transform: the circulation of
safety knowledge," in D. Nicolini, S. Gherardi and D. Yanow (eds), *Knowing in
Organizations: A Practice-based Approach*, New York: M.E. Sharpe.

Jensen, T.E. (2003), "Aktør-Netværksteori – en sociologi om kendsgerninger,
karakker og kammuslinger," *Papers in Organization*, 48.

Larsen, H.H. and C. Svabo (eds) (2002), *Fra kursus til kompetenceudvikling på
jobbet*, Copenhagen: Jurist- og Økonomforbundets Forlag.

Latour, B. (2004), "On using ANT for studying information systems: a (somewhat)
Socratic dialogue," in C. Avgerou, C. Ciborra and F. Land (eds), *The Social Study*

of Information and Communication Technology: Innovation, Actors and Contexts, Oxford and New York: Oxford University Press, pp. 62–76.

Latour, B. (2005), *Reassembling the Social: an Introduction to Actor-network-theory*, Oxford: Oxford University Press.

Law, J. (2004), *After Method: Mess in Social Science Research, International Library of Sociology*, London: Routledge.

Law, J. and J. Hassard (1999), *Actor Network Theory and After*, Oxford: Blackwell.

Nicolini, D., S. Gherardi and D. Yanow (eds) (2003), *Knowing in Organizations: A Practice-based Approach*, New York: M.E. Sharpe.

Pels, D., K. Hetherington and F. Vandenberghe (2002), "Sociality/materiality: the status of the object in social science," *Theory, Culture & Society*, **19** (5–6), 1–21.

Pratt, M.G. and A. Rafaeli (1997), "Organizational dress as a symbol of multilayered social identities," *Academy of Management Journal*, **40** (4), 862–98.

Pratt, M.G. and A. Rafaeli (2001), "Symbols as a language of organizational relationships," *Research in Organizational Behavior*, 23, 93–132.

Raelin, Joseph A. (2000), *Work-based Learning: The New Frontier of Management Development*, Upper Saddle River, NJ: Prentice Hall.

Rafaeli, A. and M.G. Pratt (2006), *Artifacts and Organization: Beyond Mere Symbolism*, London: Lawrence Erlbaum.

Spender, J.-C. (1996), "Making knowledge the basis of a dynamic theory of the firm," *Strategic Management Journal*, **17** (Winter), 45–62.

Star, S.L. (1989), "The structure of ill-structured solutions: boundary objects and heterogeneous distributed problem solving," in M. Muhns and L. Gasser (eds), *Readings in Distributed Artificial Intelligence*, vol. 2, Menlo Park, CA: Morgan Kaufman, pp. 37–54.

Strati, A. (1999), *Organization and Aesthetics*, London: Sage.

Strati, A. (2003), "Knowing in practice: aesthetic understanding and tacit knowledge," in D. Nicolini, S. Gherardi and D. Yanow (eds), *Knowing in Organizations: A Practice-based Approach*, New York: M.E. Sharpe.

Strati, A. (2006), "Organizational artifacts and the aesthetic approach," in A. Rafaeli and M.G. Pratt (eds), *Artifacts and Organizations: Beyond Mere Symbolism*, London: Lawrence Erlbaum.

Suchman, Lucy (2003), "Organizing alignment: the case of bridge-building," in D. Nicolini, S. Gherardi and D. Yanow (eds), *Knowing in Organizations: A Practice-based Approach*, New York: M.E. Sharpe.

Sundbo, J. and L. Fuglsang (eds) (2002), *Innovation as Strategic Reflexivity*, London: Routledge.

14. Intrapreneurship: differences in innovations is a matter of perspective and understanding

Hanne Westh Nicolajsen

The focus in this chapter is on the role of entrepreneurs or rather corporate entrepreneurship in the process of organizational innovation. On examining the implementation of new technology in development projects in a pharmaceutical company, we found that the networked communication technology, ProjectWeb, became a different system in the different development projects that were investigated. We argue that this is because the implementation of a new networked communication platform requires an organizational innovation process. The technology itself has little meaning in the organization. To provide value, it needs to become integrated into organizational practices and relations making it a socio-technical system.

In this organizational innovation process, the work of intrapreneurs is important as the process of integration is a matter of interpretation. Building on the concepts of sense-making and technology frames, we argue that the use of an empowerment strategy to innovation leads to different processes of sense-making providing the explanation for the different solutions found across the projects. Our study thus points to the finding that the innovation and work of entrepreneurs are not only needed in the initial phase of idea generation, but also in the phase of implementation.

ORGANIZATIONAL PROCESS INNOVATION AND ENTREPRENEURSHIP

Organizational innovation is a concept that is used in a broad sense to understand the introduction of something new, whether a new product, service or a new process. The process of organizational innovation is often described as encompassing idea generation, development and implementation (Sundbo, 1996). This definition emphasizes that something new is developed and not least that the implementation, and thus the usefulness, is important.

Kanter's definition differs from this to encompass coalition building and diffusion (Kanter, 1988). The element of coalition building emphasizes that innovation in an organization is a complex matter involving the roles of many actors such as sponsors, developers and users. This implies, first, an understanding that the intrapreneur (an organizational entrepreneur) is not an individual player acting in isolation but part of a bigger network and a system. Second, it implies a power dimension for example in terms of prioritizing the organizational resources.

The issue of diffusion implies some kind of sustainability or reproduction. According to Drejer (2004) reproduction is traditionally seen as repeating processes or reproducing products, but it may also be seen in terms of learned competencies. Diffusion is thus seen as an integral part of innovation but even in this process we may see new innovations which complicate the matter, however not all diffusion is innovative.

Some dispute surrounds the degree of newness or uniqueness which is used to distinguish between concepts like innovation on the one hand and, on the other, development and learning. Some argue that for a change to be an innovation, it has to be new to the entire market (Schumpeter, 1969 [1934]) others argue for a more local uniqueness providing added-value in a given market or to a given organization. This discussion is important, especially when dealing with customer-fitted services which could be a way of categorizing our case. Drejer (2004) argues that to the extent that these customer-fitted services are based on the combination of well-defined elements that remain unchanged no innovation is taking place. However, if new elements are developed or if well-defined elements are combined in new ways then innovation is taking place. She thus argues that for something to be an innovation, broader market knowledge has to be developed and not only organization specific knowledge.

As already mentioned we argue that innovation is taking place in our case of technology implementation as a new socio-technical system needs to be invented around the technology to provide for effective use of the new technology. We thus translate the distinction into a difference in solutions and knowledge about the matter: (a) on a project specific level; or (b) on an organizational level. However, the process of implementation is interesting no matter which level it applies to.

Intrapreneurship in the Implementation of Networked Communications Technologies

We see organizational innovation as happening within an organization's ecological system, this means that the appropriation of new technology is difficult to achieve because information ecologies are diverse, continually evolving, and "marked by strong interrelationships and dependencies among the different parts" (Nardi and O'Day, 1999). This implies a systemic understanding of innovation including what Freeman refers to as "flows to and from sources of scientific and technical knowledge as well as flows to and from users of products or processes" (Freeman, 1996). Attempts to introduce new communication technologies in organizations often fail because managers and technologists underestimate the time and effort it takes to successfully appropriate and incorporate a new communications medium into the existing "information ecology," that is, the system of people, practices, genres, and information and communication technologies in the local environment (Nicolajsen and Bansler, 2007). This points to another insight about innovation which is that changes are often found in the diffusion of innovations (Rosenberg, 1976). The findings above mean that innovation processes are much more complex and messy than what is often described in the literature. They interweave many elements in non-linear processes including the difficulties of defining the beginning and the end of these processes. In this messy process we understand intrapreneurship or corporate entrepreneurship as the work of change agents trying to create some structures that make the new technology make sense within the ecological system.

Giddens (1984) has tried to account for the relationship between individual actions and systems by arguing for a duality between individual actions and social structures as constituting one another. These structures guide us without being deterministic. The same applies to new technology, it might guide certain actions without determining them and can be seen as an occasion for restructuring the organization (Barley, 1986). When individuals such as entrepreneurs act, these acts are influenced by the structures surrounding them. By acting they either re-establish the existing structure or challenge it, which might lead to new structures, the latter case is defined as innovation. Individuals are thus seen as important change agents in any structure.

In the early theory on entrepreneurs it was argued that entrepreneurs are the ones seeing and seizing opportunities for change. Schumpeter (1969 [1934]) argued that this happens in opposition to the current situation through what he terms "destructive creation." Kirzner (1973) argued for a less drastic situation that entrepreneurship may happen in response to unexploited opportunities in the current situation. However, both view the entrepreneur as individual actors that can freely act to establish his or her own enterprise (Sundbo, 1996). In our systemic approach we would rather argue that entrepreneurs like any other agent act within a social framework influenced by others in addition to the agent's own aspirations and competences.

The individual activities and thus those of intrapreneurs can be understood through the concept of sense-making. Weick argues that each individual constantly seeks to make sense of their surroundings. The process of sense-making depends on our former history and experiences guiding our understanding and interpretation of the present situation and possibilities at hand along with a wish to create a positive identity of ourselves (Weick, 1995; 2001). In the case of organizational sense-making, this happens as a game of ongoing negotiations, repetitions and a request for people to account for and rationalize their actions. The process of implementing new technology can thus be seen as an organizational sense-making process dependent on a lot of individual sense-making processes. Orlikowski and Gash (1994) have examined the processes of how people make sense of technology. They argue that this sense-making process builds on what they call technology frames. Individual "technology frames" are ones contextual understanding of technology made up by a combination of:

1. the nature of technology, referring to people's understanding of the capabilities and functionalities of a given technology;
2. technology strategy denoting the reason why the technology was introduced and how it is supposed to create value in the organization; and
3. technology in use is the understanding of how the technology will be used on a daily basis and how it affects the users in different ways. (Ibid: 183–4).

The understanding of functionality points to the fact that it is not what the technology provides, but what is seen as possible, which again depends on what is introduced and seen and how one understands the technology by trying out and interacting with the technology in its context. The understanding of technology can be provided through means of demonstrations, user manuals and training. The strategy dimension denotes the possibility of being influenced about how to use the technology through visions and guidelines given that they may both direct use and motivate/de-motivate use. Technology in use refers to experiences with technologies. This can be experiences using other similar kinds of technologies or more broadly. But it may also be the experiences gained from using the new technology in terms of adopting or creating new practices and learning to see some of the possibilities and limitations informing future visions and use. The definition of Technology Frames implies a basis for learning through experiences, meaning technology frames are changed as experiences are gained either by oneself (technology in use) or learned from others (nature and strategy of technology) (Nicolajsen, 2005) or expanded (Bjoern *et al.*, 2006). All three dimensions are of course interrelated. The theory also emphasizes that people within certain areas have common or similar technology frames. Working with new technologies might produce some discrepancy between visions and wishes on the one hand and what is actually possible on the other, which again emphasizes that new technology needs to be learned and experienced. As experiences are gained, the actual use and strategies should converge toward one another.

Making sense of technology is a highly demanding process. To gain the most from new technology, there is a need to follow its introduction with changes in the organization, such as practices, as well as making organization specific changes to the technology, thus creating a process of mutual adjustment between technology and organization (Leonard-Barton, 1988). This organizational sense-making process is even more warranted when implementing new technologies that are what Weick *et al.* (1990) call equivoque. Equivoque technologies are open to several interpretations leading to uncertain and complex uses as well as possible misunderstandings. In the case of networked communication technologies, such as ProjectWeb, these are found to be open and complex systems and a number of researchers have discovered that their implementation is not straightforward, but requires the development of common conventions on use (Mark, 2002). Employing the terms of organizational sense-making, some navigation points which provide a common translation and organizational sense-making are needed (Weick, 2001). This process can be long and enduring due to the domino effects in the information ecology of the organization.

The initial attempts of integrating the technology easily result in unforeseen consequences requiring adjustments. Furthermore, some studies suggest that the additional work that needs to be done to make these kinds of systems succeed have most often to be provided by someone other than the ones actually gaining the advantages (Grudin and Palen, 1995). This asymmetry suggests that it might be convenient to provide resources and appoint responsibility. This kind of work is what Orlikowski *et al.* (1995) coin technology mediation. Technology mediation is the process of structuring user's use by influencing their interpretations and interactions by changing the institutional context of use and by modifying the technology itself.

In other words, the first two dimensions of the users' Technology Frames can be influenced, whereas technology in use demands individuals' own experiences. The work of technology mediators also called gardnering (Nardi, 1999) or tailoring (Trigg and Boedker, 1994) requires knowledge of the technology combined with organizational knowledge as well as influence. This description has some similarities to some of the theories addressing the qualifications of entrepreneurs (Lettl *et al.*, 2006) who are said to need a combination of qualifications in technology and a given domain. This is similar to the qualifications of technology mediators. In addition, entrepreneurs are also expected to have a special drive (ibid.) or what Kirzner (1973, 1992) defines as alertness to opportunities. Whereas it is argued that mediators need to have a profound wish and joy helping others (Trigg and Boedker, 1994).

Learning from the insights above, the implementation of an innovation such as a new communication tool is not a one-way process, but rather a question of mutual adjustment between the technology, different users and the organization. In our case, product innovation is followed by process innovation and vice verse. Seeing the needs and working out these changes can be quite demanding, breaking with the existing routines and the mental schemes that lie behind them. This work is seen as innovation, demanding some kind of entrepreneurial work to become a success. The work contributes, first and foremost, to context-specific innovations but may also contribute to market based innovations to the extent that the technology template is changed (product innovation) or if some common insights or competences can be learned and used in implementation of other systems or the implementation of the same system in another context. We thus argue that at least for some types of products, the implementation process is indeed an innovative process as they are not straightforward to use but require individual and organizational sense-making.

EMPIRICAL SETTING

Our empirical case is a global pharmaceutical company Medica (a synonym) based in Northern Europe. Medica develops, produces, markets and sells their pharmaceutical products in more than 175 countries. To stay competitive in the market, it is important to constantly develop and introduce new products. A new organizational communication platform, ProjectWeb, has been developed and pushed into development projects to help communicate and coordinate more effectively within the management group.

ProjectWeb has been developed by Medica's IT department, however the innovation process started in the user environment. The invention of a more user-friendly and web-based system made it possible for more people with limited technical competences to create internal web-pages to support their work. To prevent a beginning trend of departments using substantial resources to develop their own web-based communication system, it was decided to let the IT department take over. The arguments for developing one corporate platform instead of many locally developed systems were those of keeping costs down and securing central support both in terms

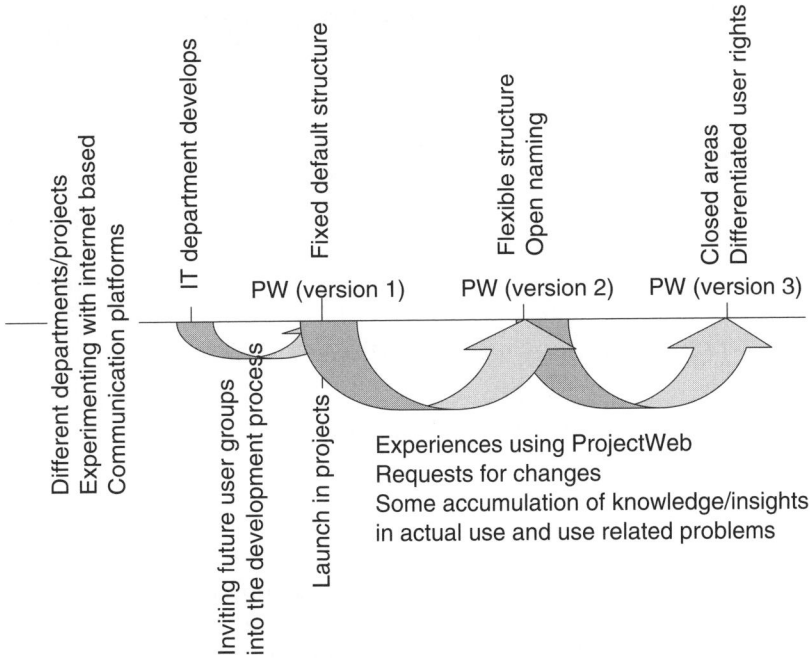

Figure 14.1   The innovation process of ProjectWeb

of updates and implementation thus saving resources on development, updating, implementation and collection of experiences. As is shown in Figure 14.1, the innovation of the template has proceeded through the phase of implementation as user-feedback requesting change is incorporated in new versions of the product. At the same time, organizational innovation processes are taking place exemplified by the two big arrows in the figure.

We started our research after the third version of ProjectWeb was launched. Investigating the implementation of the ProjectWeb template in so-called Development Projects we carried out 33 interviews with project members in different positions. We also inspected ProjetWeb in one project and collected statistics on its use along with other types of secondary material. The study was rather inductive in its approach.

Product development projects in Medica are complex, large-scale, long-term endeavors. A typical project lasts nine to ten years and involves up to 500 people from many different functional areas within the company. Most of the activities are carried out at sites in Northern Europe, but clinical trials are conducted in the US, Singapore, Japan and a number of other countries worldwide. The fact that a growing number of development projects are

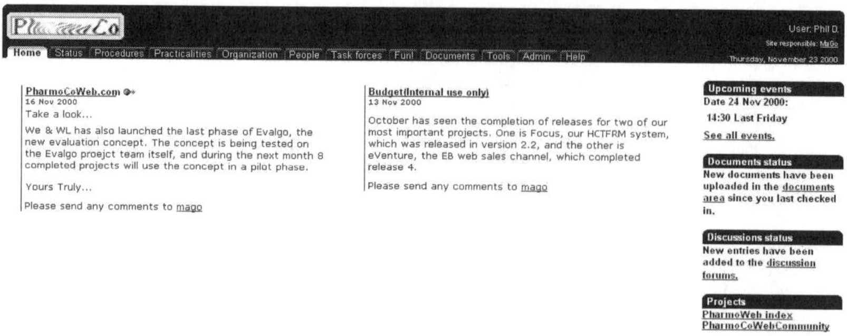

Figure 14.2   Screen dump from ProjectWeb (IT department)

undertaken in close collaboration with external partners adds to the distributed and complex nature of these projects. The development process itself is highly structured and regulated by the health authorities in various countries (see for example the Hoffman LaRoche pharmaceutical case in Ciborra (1996)).

Development projects in Medica have a standardized management structure (see Figure 14.3). On the top one finds the project director and the project assistant. Underneath come the working group leaders. Working group leaders as well as the project director and the project assistant are all members of the "core group." Usually there are around ten working groups representing different areas, for example clinical research, engineering, marketing and regulatory affairs. Project members represent employees from clerks to highly educated experts, mirroring the division of work to be found. The computer literacy of members is quite varied.

The groupware application, ProjectWeb, which we studied, is a web-based application of the virtual workspace type (similar to BSCW from GMD in Germany or Lotus Team Workplace from IBM). ProjectWeb offers facilities for sharing documents, exchanging files, publishing information, event notifications, group management and so on. It operates with four content types: events, articles, documents and discussions. The generic template allows for different forms of configuration[1] opening the possibility for material differences across projects. A default structure for projects has been developed. This default structure provides the need for simple usage, where the system administrator just needs to add users and content. The template allows for changes in the default version; each system administrator can change colors and graphics of the site and she can make changes in the content structure including renaming existing folders. In addition, users can be given additional rights to browsing; members of a project group (extended reading rights), document editor, article editor, project administrator and so

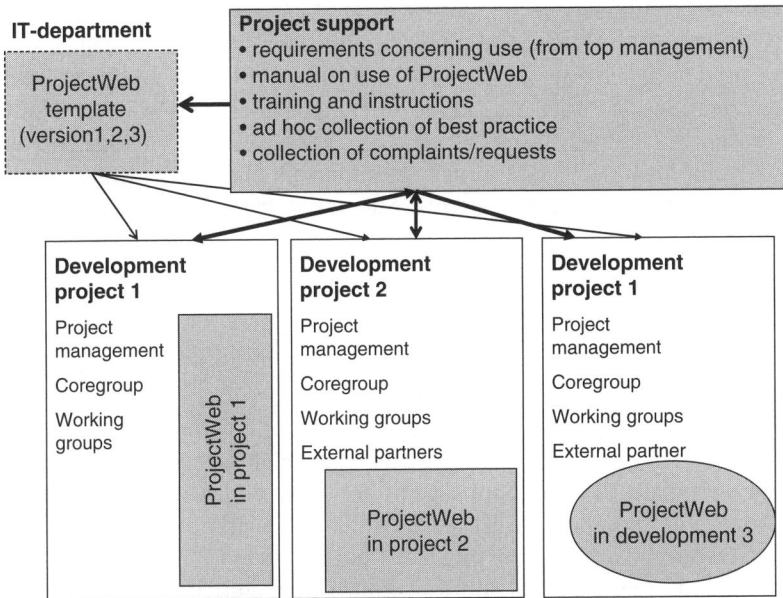

Figure 14.3   Innovation-in-use?

on. There is a restricted access function for individual documents and it is thus for the individual author to decide if restrictions are to be used (by uploading material to one of the specified groups).

The official purpose of the system is to support communication and file sharing in the Core group. At the time of inquiry, all development projects were expected to operate ProjectWeb to support communication and file sharing at least at the project level of the Core group. Project assistants in each development project were given the duty of system administrators. To help the implementation project, assistants received a general introduction to the system, they were also given an operating manual. In addition, a person with technical skills and insight into the domain of project management provided ongoing support and supervision.

The work of the support project assistant in project support can be seen as an important role in the organizational innovation system. This is because she both feeds the innovation process regarding the development of the ProjectWeb template and supports the organizational innovation process by providing the project assistants' training and diffusing solutions and best practices.

In this chapter the main focus is on organizational implementation within the development projects. In this phase of the innovation process, the company followed an empowerment strategy whereas the development phase followed an expertise strategy (Sundbo, 1996). In the figure above, we have tried to depict how this empowerment strategy leads to quite different interpretations of ProjectWeb in the different development projects. This is the case, even if the diffusion of ideas and solutions are exchanged across the development projects; through the support assistant or through interaction among the project assistants, who are all from the project management department.

Implementation of ProjectWeb

We followed the implementation of ProjectWeb for a period of three years (1999–2002). At that time, this kind of groupware system was not as widespread as it is today, this meant that there was little knowledge about how best to use these kinds of tools. There were also high expectations and hype. What we found was different usage of the same systems both across and within the development projects. The differences within the projects were expected due to variations among tasks and practices as these development projects are organizational entities comprising members in different positions and areas. The differences among development projects were more surprising as development projects have overall tasks in common and the same kind of organizational structure, though they do vary in the number of members (from 75 to 500) and whether external collaboration partners are involved or not. But these situational dissimilarities cannot explain all the varieties of use. Here, we want to give an explanation of the differences as a question of different understanding and development of the technology in use also comprising innovation-in-use.

All development projects were given a default version of ProjectWeb. A responsible system administrator was appointed and a minimum of use in terms of communication and file sharing in the Core group was requested. This was interpreted as the provision of Core group minutes! It was to a great extent up to the different projects to decide how they wanted to use Projectweb. As communication tools are seen as a service and not the core business, this is automatically a function that assistants are responsible for. What we found when entering the scene was that in many of the development projects, ProjectWeb never really got integrated in the organizational practices – only the requested Core group minutes would be uploaded, but never or rarely accessed. This pointed to weak technology frames due to lack of visions, lack of understanding of the technology and lack of understanding of how this technology might function in use preventing implementation and practical experiences to grow. When technology frames are weak, nobody is able or willing to take care of the implementation as nobody can see the value. However, in a couple of development projects, ProjectWeb flourished and was emphasized as a success. We followed three of these projects to gain knowledge about implementation and use.

The overall finding that only a few projects integrated ProjectWeb into organizational practices can be argued as being due to ProjectWeb being equivoque, it thus needs to be interpreted and translated into practice. This is not least because it is not an individual tool. As soon as it is used in group processes some common expectations and agreements on how to use the tool is necessary to make it valuable. Even in the case of limited use, there is a demand for some work to be done by the project assistant such as setting up the platform, making areas of content and entering members and so on and some limited agreements on use.

In the following, we will present the use of ProjectWeb in the three "succesful" projects along with the different processes of sense-making resulting in ProjectWeb becoming a dissimilar tool in each project. To understand the organizational sense-making taking place we emphasize a dimension of power/priority including the status and resources spent and which part of the organization is involved.

Project 1

In this project, the project assistant took on the responsibility of making ProjectWeb into a publication tool to provide everybody within the project as well as others within the organization access to most information about the project. The director and the core group supported this vision and approach. The project assistant was quite familiar with the technology and liked to experiment and usc timc to make it work. In the beginning, she mainly started to provide most of the communications she already provided. By uploading it to ProjectWeb, she made an archive of all the communications in the project and possibilities of more current updates of the material which is not possible with printed documents. The project development plan is a good example of this. The plan includes chapters about all areas in the project, and it is quite expensive to distribute. It also requires a long time to collect all the documents that were distributed meaning updates were normally delivered half yearly, and then parts of the plan would already be obsolete. She used the vision of on-line publication to break with the existing rules on confidentiality by providing far more information than is normally accessible. For example, she uploaded the Core group minutes without restriction. Other members were seldom involved and mainly limited to the delivery of requested content for the project assistant to process and upload as well as providing feedback on the content provided. A member requested a list of fax numbers however, this type of content stayed within the given paradigm and was not seen as innovative. Apart from providing information to a wider audience, she also developed new types of content as she became more confident with ProjectWeb and the functionalities and possibilities. She invented the new content type to support the creation of a feeling of community for the otherwise virtual project organization, thus expanding the initial visions.

She developed "background information" such as the story of the invention of the product along with information on current status and success-stories from around the globe on peculiar uses of a forerunner of the product. Due to the wish to use categories of content on other levels than intended and a wish to deliver new types of content this project assistant was among the ones who requested a more open structure as well as the open naming of folders. Her overall strategy was to bring the goals and knowledge of the project to the project members to provide a more holistic understanding – a big picture – which she and the project director believed would provide a background for better informed decisions across the project as well as more dedicated employees. A lot of the work she did was not straightforward – like uploading pictures. The project assistant in this project thus arrived at making a consistent and common interpretation of ProjectWeb based on her own Technology Frames. She was generally creative in the way she broke away from the initial idea of ProjectWeb to support mainly internal management communication, extending it to a publication tool from management to the periphery as well as a community supporting tool. She started out on a small scale but on the way she gained knowledge on the technology as well as understanding of how the audience can use her content. The technology frames thus changed as a better understanding of the technology in use has inspired the creation of new visions.

Project 2

The project assistant in this project also used a lot of time to make ProjectWeb work. She, like the project assistant in Project 1, started by providing plans and minutes that were earlier communicated by email or post. Over time she created new types of content, breaking with the original default structure. To support community building across the project's virtual organization, new types of communication were invented or transferred from other situations. The project assistant made room for more personal communication. An example is a funny story and pictures from a visit to Scotland where a member is wearing a kilt. This communication was provided to create a sense of knowing the others in the project, which was normally not an issue and also hard to create in these big and physically distributed projects. A huge difference between this project and Project 1 is that only project members are allowed access and restricted groups have been formed. Forming the restricted groups also required additional work by the project assistant. Moreover the idea of using ProjectWeb to support collaboration was followed by a wish to restrict access to smaller groups. Different explanations were given: to collaborate in "peace," to keep some material confidential, and to limit the amount of information. An issue was that collaboration work requires the communication of unfinished material. To support this kind of work, a need for signaling the status of the document emerged as web-tools are normally used as publication tools in Medica. To meet this need a folder for unfinished material was made. In this projectwise sense-making of ProjectWeb, the project assistant kept herself in the background opening up the possibility for broader perspectives by inviting and expecting Coregroup members to contribute to the visions and interpretations. This has meant that more visions are covered or combined in this project's interpretation of ProjectWeb. All the collaboration practices are invented within the working groups. These spread inventions are based on the project assistant providing assistance and instructions on how to use ProjectWeb for different purposes as a certain level of understanding and operating the system is necessary. By providing this kind of support she is thus stimulating the technology frames of individual users or groups of users making it possible for them to develop new types of use or to integrate already developed practices. This project assistant got additional work in coordinating different uses by creating different folders and common rules on use. This organizational sense-making around ProjectWeb thus required greater technical knowledge to support the coordination of the distributed emergence of practices. The result is a ProjectWeb that not only provides publication from the management to the periphery but also support publication, communication and collaboration across the project as well as within smaller groups across working groups as all members can publish material. These kinds of uses evolved slowly as a better understanding of the technology fostered new visions about use and as other means for collaboration such as e-mail or shared local area networks did not support the work, creating an empty space and a need. In this project, we also find examples of unfruitful experimentation as in the case of trying to coordinate translation work. However, this did not work well as the workflow and document versioning often got mixed due to lack of this kind of functionality in ProjectWeb.

Project 3

In this project, the project assistant did not have the support of her project director, who believed that other tasks were more important. As a consequence the project assistant's work in relation to ProjectWeb was rather limited and mainly directed to meet small requests from the members and the straightforward uses related to the project assistants own work; such as uploading the project development plan instead of sending it out in hard copies. The use of ProjectWeb also changed quite a lot over time in this project struggling for a purposeful and easily manageable structure of ProjectWeb. In the beginning, one ProjectWeb common to the two Coregroups was operated. The vision of the ProjectWeb was a filter between the management groups and the rest of the organizations. Later on an internal ProjectWeb was implemented covering all the project members in Medica. In the beginning, most members were just given browsing rights, except for the Coregroup who had author rights. Later on in the process, the project assistant tried to phase out the internal ProjectWeb, as it was too hard to keep both the internal and the external ProjectWeb updated. More and more of the active members were connected to the external ProjectWeb – the one that started out mainly for the Coregroups. At the end of our period of investigation, most members were connected to a group and almost all members had author rights. The "late" external ProjectWeb mainly supports communication and collaboration across the project as the project assistant and the project management play a limited role as content providers. It is thus up to the working group leaders to establish a vision as well as practices concerning use. Even they do not get much guidance, they build up some practices. The problem of this distributed approach is that some uses clash as ProjectWeb is structured to support one use only without thinking about the needs in other uses. This points to the need for some overall balancing of the different needs and practices. A concrete example of this challenge is the communication of working group minutes which clash with collaboration and the exchange of drafts. The working group minutes are not communicated openly via ProjectWeb, but to restricted groups. The restricted groups are basically defined as the working groups. However, to support coordination across the project, there is a need to support the distribution of these minutes reporting on status, challenges and future work to members in other working groups. This means that most working groups have been extended to cover additional members from other working groups. The extended working groups have exposed a problem in the way the use of drafts is structured. To support collaboration, a folder for drafts has been made. All drafts are uploaded in one common drafts folder. The use of restricted groups is a way of limiting information through access. However, for some working group members, this system does not work well as some are members of a couple of restricted groups (to get the meeting minutes), the consequence is that the content in the drafts folder is exploding and the need to keep unfinished material close is not met. The problem is minor but does require a coordinator for its solution. The project assistant has been informed about the problem, but she does not know how to solve the problem, and she does not feel she has got the resources to take care, as a result nothing is being done. The unresolved challenges using ProjectWeb in this project points to lack of common strategies and visions. The explanation is a combination of lacking resources and qualifications, and nobody being able to take the overall responsibility. In our case it leads to too many different, clashing, technology frames. There is thus a lack of collective sense-making around ProjectWeb in this project. Technology frames are also changing in this project as experiences are gained, but too many disconnected activities are taking place resulting in a mess.

Intrapreneurship in the Implementation of ProjectWeb

Table 14.1 presents some of the factors which are important to the sense-making process that is taking place in the implementation of ProjectWeb in the development projects. We found that in this empowered implementation process, people need a certain level of technical understanding of ProjectWeb to be able to see the opportunities of its use. A first step toward implementation is appointment of a knowledgeable person with flair and interest in these kinds of communication technologies. Either they will provide content and functions on their own, or they will need to help others gain the necessary knowledge to support technology frames strong enough to create new practices. They will also need to manage a coherent structure

Table 14.1 ProjectWeb content and sense-making conditions across projects

	Project 1	Project 2	Project 3
Developed Content/ function	Publication (C) Community building	Publication (C/L) Community building Collaboration	Publication (C/L) Collaboration
Resources to make sense of ProjectWeb	Yes	Yes	No
Coordinating sense-maker	Project assistant	Project assistant	None (project assistant)
Sense-maker	Project assistant	Core group	Open (everybody)

Note: C = central L = local.

for ProjectWeb. If more people are involved in the sense-making, it means integrating and balancing different inventions, which seems to be quite demanding. We also found that when more people are invited into the sense-making of ProjectWeb, more perspectives and a more valuable innovation may be created. We argue that this is because it is very difficult to actually present other perspectives. However, inviting more people to innovate in relation to the use of ProjectWeb requires a central coordinator to balance and integrate different uses into the structure of ProjectWeb. As the project assistants are appointed as system administrators, they are automatically appointed this coordinating function. If they do not get the resources or if they do not have the qualifications or interest in the developing qualifications, it is likely that ProjectWeb will not be successfully implemented.

The three projects thus exemplify three different situations relating to how the implementation of ProjectWeb proceeds very differently according to the resources giving to the project assistant, the technological flair of the project assistant, and the degree to which others are invited into the organizational interpretation of ProjectWeb. In Project 1, a common and very coherent interpretation of ProjectWeb was worked out as it was solely based on the technology frames of the project assistant. However, despite this narrow approach being very ambitious and competent she made ProjectWeb into a communication tool that was appreciated. In Project 2, they succeed in creating a common understanding based on a number of technology frames. Consequently, ProjectWeb became integrated into daily and local practices along with management communication. The result is a communication tool meeting the needs of more people, as the project assistant nurtured and integrated different sets of technology frames. Her role is essential in coordinating this organizational sense-making process, building a common interpretation comprising more visions and support aligned technology frames. In Project 3 more people are involved in forming ProjectWeb according to their individual technology frames. However, there is a lack of coordination resulting in breakdowns and inefficiency. Thus there is a lack of an organizational sense-making process to align the different technology frames or balance them.

DISCUSSION

Based on our findings, we argue that organizational innovation, as in the case of technology implementation, is individually driven based on individual technology frames acting as the basis for making sense of ProjectWeb, as is the case when new practices are developed. To create innovations meeting the needs of different groups there is a need to open up for broad intrapreneurship (as in Projects 2 and 3), however, distributed intrapreneurship does require management, here seen in the form of a coordinator to support an organizational sense-making process to inform and balance individual ones.

If a company chooses to use the empowerment strategy in the implementation of new communication technology, they should consider how they can support this strategy. First of all they need to appoint a technologically competent person as system administrator and technology mediator to ensure an organizational sense-making process to guide the individual sense-making. To further nurture and guide the sense-making efforts related to the implementation of ProjectWeb an innovation department (Sundbo, 1996) can be created in order to accumulate knowledge gained from the empowered efforts to learn more widely from local experiences and support future implementation and use. The work of project support has some similarities to an innovation department, although in our case it is limited to internal diffusion in Medica and the extension of an innovation in a very limited area. But the function of collecting valuable knowledge and ideas is the same. In Medica it happened occasionally. It could have been more formalized by making ProjectWeb a subject discussed at the project assistants' meetings, or by codifying different ways of using ProjectWeb, making the learned easier to transfer.

Our study points to the fact that much creativity and a lot of work is needed throughout the process of implementation implying that the idea of entrepreneurship has to cover processes which extend products or services. Furthermore, much work is needed for the creation of innovative solutions to make the innovative service or product succeed in the specific context of use. An empowerment strategy concerning new communications technology might be a good option as the potentials of this kind of technology are poorly understood.

We found a specific need for entrepreneurs to pay attention to the organizational innovation related to the implementation of a product innovation. Innovations are invented not just in the beginning of the innovation process when a new product is developed, but also later. Despite the focus on the individual entrepreneur, the dependence on other people's involvement has been emphasized in referring to roles such as the capitalist (Schumpeter) the sponsor, developer and so on (Kanter, 1988). However, many of the roles identified by Kanter are not only developing or implementing the current idea, but expand on it and thereby contribute to innovation. Based on our case material, we have exemplified situations where the work of the "supporting" employee may be categorized as that of intrapreneurship. We find support for this argument in the literature on customer and user involvement. Furthermore, we want to argue that the insights gained in practice are highly relevant to further innovation concerning new technology, and are a reason why an open innovation process between producer and consumers is meaningful as already argued by a number of studies of innovation (Chesbrough, 2003; Von Hippel, 2005). Our study, however, shows that this also applies in an organizational setting.

CONCLUSION

A major finding in our study is the specific need for entrepreneurs to ensure the organizational innovation related to the implementation of a product innovation such as technology. Our study exemplifies that innovations take place, not just in the beginning of an innovation process when a new product is developed, but may also happen later as in the implementation adjusting the technology to the organization and vice versa. Another insight is that much of the innovation around such technologies depends on the understanding of the technology in use pointing at a need to allow for ongoing innovation based on the experienced understanding of the contextual potentials.

Our findings show that using an empowerment strategy of exploring the organizational possibilities of new technologies may be a good strategy allowing for a great deal of creativity. Such a strategy opens up the possibility for innovation in the implementation as individuals acting as entrepreneurs will develop different uses of new technology to support their own work. However, our study also shows that if nobody has the initial faith in and understanding of the technology little will happen, as nobody will be able to see or seize the opportunities provided. However, to the extent that the qualifications and the drive are there, different practices will start to develop further nurturing the understanding of others, which might lead to new processes of innovation. We also found that it is extremely important to make sure that at least one central person processes the right qualifications and interests to ensure integration either by doing the work on their own or by coordinating others' inventions. Finally, the study shows that innovations are formed by individual contributions. Having just one person forming the technology results in a limited but easily controllable technology. Whereas inviting more people to innovate and form the technology provides more perspectives and possibilities of a more elaborated innovation, however this solution demands management in terms of coordination and creative solutions to integrate different practices in the same technology.

The chapter has revealed important issues related to the concept of care as defined in the introduction of this book. First of all, it has shown how the sense-making process in innovation is based on individual experiences and understanding, but that this understanding is grounded in social practices and can be nurtured and guided through experiences, training and visions. Our study shows that in the innovation process, the involvement of the users is very important. User representation is difficult due to the differences in technology frames limiting the understanding of other needs, including what is important to others. At the same time as the innovation process is fed by individual contributions, there is a need for positioning, which is relating the individual needs to those of the organization through organizational sense-making processes providing some frames or prioritizing different individual needs.

NOTE

1. Configuration in terms of the different choices inherent in the template is almost left out here as it merely complicates the matter without providing further to the conclusions.

REFERENCES

Barley, S.R. (1986), "Technology as an occasion for structuring: evidence from observations of CT scanners and the social order of rediology departments," *Administrative Science Quarterly*, 31, 78–108.

Bjoern, P., A. Scoupola, A. and B. Fitzgerald (2006), "Expanding technological frames towards mediated collaboration," *Scandinavian Journal of Information Systems*, **18** (2), 3–42.

Chesbrough, H. (2003), *Open Innovation*, Boston, MA: Harvard Business School Publishing.

Ciborra, C. (1996), "Mission critical: challenges for groupware in a pharmaceutical company," in C. Ciborra (ed.), *Groupware and Teamwork*, New York: John Wiley & Sons, pp. 91–120.

Drejer, I. (2004), "Identifying innovation in surveys of services: a Schumpeterian perspective," *Research Policy*, **33** (3), 551–62.

Freeman, C. (1996), "The greening of technology and models of innovation," *Technological Forecasting and Social Change*, **53** (1), 27–39.

Giddens, A. (1984), *The Constitution of Society. Outline of the Theory of Structuration*, Berkeley, CA: University of California Press.

Grudin, J. and L. Palen (1995), "Why groupware succeeds: discretion or mandate?" in *Proceedings of the European Conference in Computer Supported Cooperative Work*, Boston, MA: Kluwer, pp. 263–78.

Kanter, R.M. (1988), "When 1000 flowers bloom: structural, collective, and social conditions for innovation in organizations," *Research in Organizational Behaviour*, 10, 169–211.

Kirzner, I.M. (1973), *Competition and Entrepreneurship*, Chicago, IL: The University of Chicago Press.

Kirzner, I.M. (1992), *The Meaning of Market Process: Essays in the Development of Modern Austrian Economics*, London: Routledge.

Leonard-Barton, D. (1988), "Implementation as mutual adaption of technology and organization," *Research Policy*, 17, 251–67.

Lettl, C., C. Herstatt and H.G. Gemeunden (2006), "Learning from users of radical innovation," *International Journal of Technology Management*, **33** (1), 25–44.

Mark, G. (2002), "Conventions and commitments in distributed CSCW groups," *Computer Supported Cooperative Work*, 11, 349–87.

Nardi, B.A. (1999), "Cultivating gardeners: the importance of homegrown expertise," in B.A. Nardi and V. O'Day (eds), *Information Ecologies: Using Technology with Heart*, Cambridge, MA: Massachusetts Institute of Technology, pp. 139–51.

Nardi, B.A. and V. O'Day (1999), *Information Ecologies: Using Technology with Heart*, Cambridge, MA: MIT Press.

Nicolajsen, H.W. (2005), *Tilpasning af groupware i organisationer – betydningen af metastrukturering*, Lyngby: Technical University of Denmark.

Nicolajsen, H.W. and J. Bansler (2007), "Evolving information ecologies: the appropriation of new media in organizations," in S. Heilsen and S. Siggaard (eds), *Networked Technologies*, Hershey, PA: Idea Group.

Orlikowski, W. and D.C. Gash (1994), "Technological frames: making sense of information technology in organizations," *ACM Transactions on Information Systems*, **12** (2), 174–207.

Orlikowski, W., J. Yates, K. Okamura and M. Fujimoto (1995), "Shaping electronic communication: the metastructuring of technology in the context of use", *Organization Science*, **6** (4), 423–44.

Rosenberg, N. (1976), *Perspectives on Technology*, Cambridge: Cambridge University Press.

Schumpeter, J.A. (1969 [1934]), *The Theory of Economic Development. An Inquiry into Profits, Capital, Credit, Interest and the Business Cycle*, Cambridge, MA: Harvard University Press.

Sundbo, J. (1996), "The balancing of empowerment. A strategic resource based model of organizing innovation activities in service and low-tech firms", *Technovation*, **16** (8), 397–409.

Trigg, R. and S. Boedker (1994), "From implementation to design: Tayloring and the emergence of systematization in CSCW," in *Proceedings of the Conference on Computer Supported Cooperative Work*.

Von Hippel, E. (2005), *Democratizing Innovation*, Cambridge, MA: MIT Press.

Weick, K.E. (1995), *Sensemaking in Organizations*, Thousands Oaks, CA: Sage Publications.

Weick, K.E. (2001), "Sensemaking in organizations: small structures with large consequences," in K.E. Weick (ed.), *Making Sense of the Organization*, Oxford: Blackwell, pp. 5–31.

Weick, K.E., P.S. Goodman and L. Sproull (1990), "Technology as equivoque: sensemaking in new technologies," in P.S. Goodman and L. Sproull (eds), *Technology and Organizations*, San Francisco, CA: Jossey-Bass, pp. 1–44.

15. Mindful innovation

Poul Bitsch Olsen

Mindful innovation is an approach to innovation that pays attention to people's experiences in an organization rather than to formal organization or social role. Innovative organizations must deal with and try to benefit from the experiences of their employees. How can they do this?

The chapter addresses this question through a theoretical reflection on innovation and change at the micro-level. It draws on symbolist and interactionist approaches to organization and innovation. Innovation and change are understood as moments of action in an interactive process that involves action, reflective responses to action and experience.

The chapter will illustrate some of the points through a case study of a handball coach and his team during five matches. The case illustrates some of the practical difficulties associated with small changes as well as sense-making during change.

INNOVATION AND INTERSUBJECTIVITY

Innovation could be seen as an activity that involves tensions between existing worldviews, and a preparedness to reconsider experiences on the basis of new understandings of a situation. Innovation therefore requires a mindful, imaginative or careful approach that can deal with these experiences, rather than merely rationalized strategizing. The latter approach fails to catch the fact that an innovation's success is a piece of ongoing, experience-based work, rather than a lucky punch.

This chapter deals with an interpretation of innovation, which may be called "innovation as organizing." Innovation is understood as being based on careful intersubjectivity and symbolic interaction among people. To observe the basic elements of innovative actions, we have to address intersubjective situations where people agree upon significant changes, and where the organization is consequently given a new direction – this is an innovation in the organizational sense (Weick, 1995: 73; Blumer, 1969: Chapter 1).

Innovation can be understood as an activity that involves many interacting people rather than something that takes place in a laboratory or is planned by management. The term innovation refers to the process of realizing new ideas and experiences in an organizational context (see the definition by Mulgan and Albury (2003) in the introduction to this volume). Innovation is also related to the experience of disruptions, where new agreements among employees are produced. The people related to the situation where innovation takes place interact and make some sense of what the situation is about, and in this act, they structure their knowledge of the world and decide what the world is like. So innovation is organizing based on collective sensemaking about old and new experiences and how these experiences work and are retained. Innovations become innovations only if they become part of interactive and collective sensemaking processes in an organization.

The individual perspective is only one element in this organizational perspective. No individual person is able to make an innovation come through all by herself. Even an entrepreneur or a manager will always work on behalf of others, and must be recognized by them. Attention must be devoted to their working environment, their intentions and competences. Several people must join the innovative situation. They all have personal biographies, are involved in different economic and social exchanges, mutual obligations and actions.[1] They co-reflect on the innovative situation together with others, furthermore they will enact the changes in organizational processes. In this way, symbolist theories consider organizations and businesses to be produced through collective situations, where meaning has no individual source but rather is an ongoing collective synthesis.

A basic feature of the organizing approach to innovation is its understanding of experience as a dynamic category: Experiences, which can be understood as certain descriptions or "theories" of a situation (such as a handball match), do not determine action. They are provoked by certain actions and interruptions in a given situation. This can mean that the description or "theory" of the situation or practice is reconsidered. For example, a new tactic is reconsidered for the handball team.

"Practice" (for example the handball match) is the place where knowledge and all collective value is produced, reproduced or changed. Practice represents a collective knowledge process within an organization's organizing activities. Organizing is characterized by both change and stability, but the organizing process is in no way determined by experience, even if all participants have their memories and have a need to predict the other participants' actions. Innovative processes are always related to retrospective sensemaking about what happened which paves the way for new actions and new experiences that were not possible or available before. A handball match, with its changing tactics during a game, is a good illustration of this.

VIRTUE AND SELECTION

When a handball team introduces an aggressive flat defence into its play instead of a more offensive defence, all the players have to learn about the implications of this for a wide variety of situations. It is complicated to a degree that it is sometimes almost impossible to fully understand. This means that most teams try to stick to one type of defence, and make this type the most dominant in their play. In a game, the better team has the power to decide which tactic seems to be the best and to change accordingly.

In a work organization, innovation implies that many small changes are implemented with more or less ease. Large changes that happen with little ease can leave the entire work organization untouched, like the handball team that cannot change its tactics radically. Many small changes, which are implemented with care and energy, will eventually change the organization. These changes may improve the quality of the work process. Each separate step, which demands attention on behalf of the participants, may be a reason for improvements in quality. The quality of an innovation relies on the sense-making processes that take place during the organizational practice.

How can participants evaluate where the higher value and better quality lie within their practice? Alistaire MacIntyre's definition of virtue may help us to improve our understanding of value and quality from an experience-based perspective. Virtue is, according to MacIntyre, the capability to achieve goods: "A virtue is an acquired human quality the possession and exercise of which tends to enable us to achieve those goods which are internal to practices and the lack of which effectively prevents us from achieving any such goods" (1981: 191).

Virtue is the capability that leads practitioners to achieve the internal goods inherent to a practice, for example the practice of playing handball in which one can acquire certain tactical and strategic skills. Internal goods are opposed to external goods such as money or fame. Virtue may consequently be an important topic in the study of quality and innovation from an experience-based perspective, because it supplies us with a notion of an inherent quality in a particular field of practice that is worthwhile striving for. Virtue is a driver of innovation.

MacIntyre (1981: 187) defines practice in a way which is consistent with the understanding of practice within the symbolic-interactionist perspective in which this chapter is rooted. He writes:

> By "practice" I am going to mean any coherent and complex form of socially established cooperative human activity through which goods internal to that form of activity are realized in the course of trying to achieve those standards of excellence which are appropriate to, and partially definitive of, that form of

activity, with the result that human powers to achieve excellence, and human conceptions of the ends and goods involved, are systematically extended. Tic-tac-toe is not an example of practice in this sense, nor throwing a football with skill; but the game of football is, and so is chess. Bricklaying is not a practice; architecture is.

The particular virtue describes the ability to develop the practice and to extend the human conceptions of that practice, just like a violin player can extend the practice of playing the violin. The benefit of innovation can, along these lines, be the achievement of an internal quality or standard of excellence – in addition to fame and money.

Innovation also involves an act of selection. This means, according to the evolutionary theory of organizing (Weick, 1979), that the meaning of an act is selected. This meaning is retained and thereby eventually noticed as a different experience. Selected experiences are shared with other partici-pants through a process of retention and collective memorizing.

The skills to deal with new experiences differ between organizing groups and participants, which will be characterized by different innovative pro-cesses. They become alert about different outcomes of actions and respond to them in different ways. So, one may expect to find different capabilities to deal with innovative processes in different groups, expressing different abilities to select and retain new experiences in a practice.

INQUIRY

The approach to innovation presented here continues, to some extent, Dewey's "ethical postulate": "In the realization of individuality there is found also the needed realization of some community of persons of which the individual is a member; and, conversely, the agent who duly satisfies the community in which he shares, by the same conduct satisfies himself" (Dewey, Early Works 3: 322).

Dewey emphasizes the agents' co-production of community, and that the active agent experiences intensity by satisfying the community. One could imagine that the agent enjoys the innovation process because it is a process of change where more relevant work and production is the outcome.

The production of society can be seen as a history of attempts to change and stabilize processes that are theorized on the basis of noticed unex-pected outcomes. Further, Dewey defines inquiry as an act of innovation: "Inquiry is the controlled or directed transformation of an indeterminate situation into one that is so determinate in its constituents distinctions and relations as to convert the elements of the original situation into a unified whole" (Later Works: 108).

The "inquiry" is the urge to transform the situations into innovations. Inquiry helps us to notice significant situations and actions, and to contribute to collective action. As for Dewey, innovation referred to major man made changes in the practical world. It results from ecological occurrences, deliberate experimenting and alert theorizing about the organized world.

SAFETY AND CARE

During the 1970s, Barry Turner initiated a theoretical discussion about the sociology of disasters (Turner, 1978), and, later, a discussion about care and carelessness (Turner, 1992) as elements in organizing safety. Along with Charles Perrow's contributions, these views opposed mainstream rationalist organization theories, which studied the advantages of well planned and structured organizations, such as nuclear carriers, chemical plants and power supplies.

In relation to the innovation aspects of mindful organizing, we should recall how the idea of safety culture is opposed to the idea of the potential disastrous consequences of technological changes and, in particular, to the possible implications of tightly coupled organizational structures that do not allow for learning.

There were several people who took part in these discussions, but two are especially important. Karl Weick developed a theory of risk preventing or risk enabling culture (Weick, 1987). Charles Perrow (1984) formulated the natural accident theory. He argued that the organizational and technological structure might, despite being the best-prepared social system, be without the ability to prevent disasters and accidents.[2]

A particularly important aspect was the attempt to emphasize how care for information relevant to accidents might have a critical role – the ability to learn, to make better sense and to know better. Also of importance for understanding innovation is the emphasis on the implication of care for situations that are unexpected and difficult to understand.

Care is often absent in relationships that have many standards, rules, decisions and habits. The participants are supposed to reproduce the existing rules. By contrast, the presence of care means that observant and mindful participants can respond to and learn from data.

MINDFULNESS

Mindfulness is an element in a so-called materialist epistemology, where knowledge is seen as the outcome of action that is noticed and made intelligible. Mindfulness is the moment of attention to remarkable out-comes of action. It is the attention given to provocations to existing narra-tives, theories, strategies and frames that represent the history of a person or a work-process in context. Weick and Sutcliffe define mindfulness in the following way:

> By mindfulness we mean the combination of ongoing scrutiny of existing expec-tations, continuous refinement and differentiation of expectations based on newer experiences, willingness and capability to invent new expectations that make sense of unprecedented events, a more nuanced appreciation of context and ways to deal with it, and identification of new dimensions of context that improve fore-sight and current functioning. (Weick and Sutcliffe, 2001: 42; 2006: 516)

The authors give credit to Ellen Langer and social psychology for this definition. But it is also related to the evolutionary organizing model as defined be Weick (1979). In that model, change is described in terms of the enactment, selection and retention of meaning. Selection is about collec-tive mindfulness. This also connects to the sensemaking-perspective devel-oped by Weick (1995) where the process of sensemaking is activated by the participants' ability to initiate intersubjective negotiations about experi-ences that are different from what was expected.

There is an important difference here between sociology and social psychology. In sociological analyses of organizing we are concerned about

---

BOX 15.1   DEFINITION OF MINDFULNESS (FROM WEICK AND SUTCLIFFE 2001 AND 2006)

Mindfulness can be defined as a combination of the following elements:

- Ongoing scrutiny of existing expectations.
- Continuous refinement and differentiation of expectations based on newer experiences.
- Willingness and capability to invent new expectations that make sense of unprecedented events.
- A more nuanced appreciation of context and ways to deal with it.
- Identification of new dimensions of context that improve foresight and current functioning.

the interactions with the collective subject (the "action net" or group of participants in John Dewey's and David Silverman's terms (Silverman, 1970: 151)). Social psychology insists on the study of personal identity, personality being seen as a response to the process of intersubjectivity.

THE COACH

In the following, we discuss an example of a handball coach in the male handball league during five matches. Team sports constitute a practice of continuous micro-innovation in which success is measured after one or a few hours of play. After each match we know whether the changes made before and during the game were appropriate. The purpose of innovation is to win, and to prepare the team for future matches.

The example below concerns a coach (NM) who at the time had been a league and division coach for 15 years. Previously he had been a top league and national player. He works as a schoolteacher, which allows him to change his work duties at relatively short notice.

After each match, the handball league coach is in the situation where he can be alert and mindful over the various new experiences he has just had. When a match is finished, it is followed by two days when the players and the coach relive the processes. This is possibly a process of rejection or forgetting things or reliving successes. After the two days everybody must be ready to continue training and preparing for the next match.

The coach should not deny any experience that is related to the game. On the contrary, the coach has a role that is based on participating in and contributing to the teams' sensemaking and to the individual players' sensemaking. This is done in dialogue. So he must prepare himself for the next training session and being with the players again.

Making a change has both an abstract side, because it must be theorized about when it is discussed, and a practical side, which includes the possibility of improving concrete action. Furthermore, it must be integrated into many different types of situations, when the ball is played around during a match.

For both coach and players, mindfulness is an important principle, and it represents for the coach an important tool for him to stay in the teams' collective process of sensemaking. He achieves this by re-establishing himself in the new, emotional, situations in which the players see themselves based in the recent experiences in a match.

The information about this coach's participation in "mindful innovation" was gathered in between and after five matches. An interview was held between two to nine days after each match. As a coach is a busy person with a full time schedule, the interviews were done when it seemed possible, usually one or two days before the next match.[3] The interviews were video-taped and analysed.

The series of matches was well chosen, but the timing could have been improved so that interviews could have taken place when the trainer was more apt to reflect on his new experiences. Observation and following the coach as a shadow would have been an even better strategy to notice mindfulness, to be there when recognition of new experiences takes place and becomes present for him.

The advantage of the present material is that it is quite robust. It includes mindfulness that has had some significance for the coach before the upcoming matches, and most interviews took place while the coach was quite rested and motivated for discussions. The coach reported about what happened during the last match, and what he noticed and what still appeared relevant for him about that experience. So, to a large extent, the interviews express the mindfulness as it was reported by this coach during the process of change. The interviews were made in a situation where the consequences were not known, and the relevance of the response not yet tested.

The material is consequently a report from a sensemaking process. It may be compared to similar situations in other organizations when people try to incorporate experiences of their collective performance into an uncertain process of improvement, inquiry and mindful innovation.

In this case, there are some unusual circumstances, which make it especially interesting as a case of mindful innovation. The team, which is investigated, is a new league team which must face the insecure status of being in the league with players not all of which are prepared to meet the required standards, and whose qualities are important, but not sufficient to support the team's ambitions which are contested by better teams. The teams they played against were: GOG (stronger and better players), Kolding (players who were able to vary their play during the match), Viborg (swift and technical players with better technical potential), Lemvig (players of the same standard as our team), Bjerringbro (skilled and much stronger players that had had great success in matches at that time).

A source of experience was that the team did much better against the good teams (GOG, Kolding and Viborg) than expected, even if they lost all three. The coach's fear of humiliation in these matches was used by the coach to turn the situation around: the team could use the experience of these matches to change the image of the team. Also, he drew upon the lesson gained from the matches with GOG and Kolding, namely that the team must practice much more in terms of physical and powerful attack-defence situations, so they would get use to the level of violence and the standards in the league – these are, in a sense, the internal goods that must be uncovered and achieved.

The concept of mindfulness is related to the coach's general understanding of the matches: the first three were, as mentioned, against very good teams, where the survival of the team's participation in the match was more important than gaining a point. He used all the players against GOG so that they could learn what the level in the league was like. This was the intention as well against Kolding and Viborg, but he changed the strategy slightly, as he saw "possibilities for a miracle", particularly in the Viborg-match, which made him select the best players for this battle.

When the three matches were over, a home match against Lemvig had to be won. The coach prepared himself and the players for a very physical match, using all the best players. The complete process of the match was discussed. At half time, the coach was concerned because the players – in his opinion – were too passive and lacked initiative. But the match was won with a narrow margin. It was in his opinion won due to the initiatives he took in the second half.

The last match was against Bjerringbro, which were not far ahead of them in the league. But the coach prepared the players to be alert and to deal with opponents having more skill and physical strength and spirit than the newspapers had reported. Nevertheless, this match was lost. After this defeat, the coach was in no way surprised about anything, but looked forward to a long break – which is unusual for this team – and to three matches from which the team needed to gain some points to maintain a better hope of staying in the league (only one point was gained from these three matches, which we will not deal with here).

During the break, the coach revised his assessment of the team. At this point he saw its potential to be amongst the two or three minor teams in the league, not among the four or five lowest teams, which he had said in the beginning of November. The last match against Bjerringbro told him that the players' understanding of the game and mentality were sufficient for being in the middle of the league, so training and tactics had the right quality, but there was a need to develop the technical and physical abilities of the individual players.

The coach gained some new experiences. That is, some players showed new qualities, in particular during the matches against the best teams. These experiences gave the coach new ideas about the collective play and tactics, which he can use if the game develops as he imagines.

The coach also experienced unexpected possibilities through the players' mental response to situations. The spirit of the team and its capabilities were not easy for the coach to be precise about in advance. It was something that had to be noticed and specified while it happened. Maybe it could be responded to through changing the tactics. Mindful innovation, in this case, was something that was carefully and imaginatively unfolded during the matches through action, selection and retention of experience.

Looking at mindfulness in particular, some comments may be made about what the concept meant for innovation. It was about the experience in the team of the internal goods that could be achieved by extending the practice of playing handball; it was about building the virtues or capabilities that are necessary for achieving this goal.

In the following, the various characteristics of mindfulness (see Box 15.1) in relation to the case study and innovation are briefly explored.

Ongoing Scrutiny of Existing Expectations

During the first interview, NM mentioned a new left wing that went into the match and scored three remarkable goals, while the established wing had a less successful time. Nevertheless, two weeks after, this new person was out of the team. The player was told to stop because he "acted like a seventh grade child."

Also during an interview, NM stated that he saw new qualities among the players that he had not foreseen, and therefore he had to respond to them in a way that is not standard procedure. The collective experience of a new quality had to be expressed, exploited and pursued to some extent.

Ongoing scrutiny has its limits because the coach must also build and adjust expectations. So the team visualized possible situations in the play so as to produce new collective visions of the team and its abilities.

Refinement and Differentiation of Expectations Based on Newer Experiences

The differentiation of expectations can be understood as preparedness to accept the results of the game, after having prepared the team to win or do well. NM expressed, several times, a preparedness to accept the results, and at the same time a commitment to winning the games. For this team, it means that the collective ideology, which says that winning is necessary, is embodied in the players in the sense that they are happy about any victory and sometimes satisfied with a promising defeat.

The coach tries to see fights with superior teams as possible mental disasters for the team, which must be avoided by reformulating the ideology of the commitment to win. He told the team that for a specific team, winning is about staying in the league. In the first half of the season, it was unclear whether the team's aim should be to try to avoid extra qualification matches, but at a later stage they knew that this was no longer relevant.

NM responds to most situations and the experiences he gains from these as if they were included in the expected universe. Still he tries to make sense of what happens, and to reuse it in the preparation for the next game. He reformulates the events into strategies as if the strategies had been there for some time. He has an idea of what can happen, and uses this idea in the presentations for and discussions with the players.

Games that are lost beforehand, may call for strategies that will provide the team with excuses. This was the case in the first two games when he expected to lose (GOG and Kolding), and he used all the players including the new and untested. Refinement is possibly a better formulation of the strategy.

After four very good matches, NM had a hope of taking one point from one of the leading teams (Viborg), and only used the established players. NM also explained his scepticism about the possibility of winning the fifth match, against another weak team, and reasoned that this team had also shown progress in form, not having lost the five last matches.

Willingness and Capability to Invent New Expectations that make Sense of Unprecedented Events

The invention of possibilities against a top-team and the de-formulation of expectations about the last match are examples of how NM changes the picture against the background of recent experiences. It seems easy for him, because earlier expectations had been fulfilled. In a way, his ability to describe the situations is very good at a general level. It seems as if he is able to know a reality before it occurs. He tries to explain components and elements of new experiences that are visible and seem of significance for the results.

A More Nuanced Appreciation of Context and Ways to Deal with it

NM is the first to explain the match in its context for journalists (he spends time on the telephone with the two regional papers), for the players (in the ongoing collective sensemaking process), club members (he expresses his opinion of organizational matters) and the sponsors (he is always explaining the match to the VIP-group before matches). This appreciation of the context has the consequence that the team as a whole may gain certain new expectations about the future.

It has so far been difficult to come closer to NM's considerations about these things, which might be possible under better conditions for dialogue and observation. The particular situation for a coach is that he is both the strategizing and the legitimating person when press, managers and sponsors are present. Doubt about the strategies are not presented front stage, but belong to back-stage activity, and are only presented in retrospect.

New Dimensions of Context that Improve Foresight and Current Functioning

NM's formulation of the individual player's competences compared to competitors' players and their expected ability to fight the more advanced players' game, become stronger over time, and the strategies for the players were better formulated during the last interview. The context had revealed itself for him, in the sense that some aspects were no longer open to doubt. An example is the idea that some of the players are sufficiently competent; now they need more experience and training most of all. But he is also more precise in his statement that the team needs members that can move some attention away from important players, in particular it needs large (tall, strong, fast, heavy) and technically skilled players.

Mindfulness is thus a concept that explains important parts of the strategizing process and the changes that lead the participants to a new situation. The trick is to reformulate the team's resources and developments, things that are within the powers of the team to change. A critical point is definitely that the coach sees himself as the person that should know the potential of everybody. The coach aims to stay in touch with – and be responsive to – all the experiences the team has. Furthermore, he has to recognize the resources and suggest possibilities for development change. This happens during the week, when there is intensive contact with the players. And it happens during the match.

Not in all situations of mindfulness would one also talk about innovation, rather one should talk about relevant organizing. But the mindful response to experience would quite often include changes that come under the definition of innovation.

Collective Mindfulness

When the coach sees signs of some weakness that can be of significance for the team, he discusses them with the team. He continues to work on the meaning of these signs of weakness, and on what can be done. This reflection addresses "the course of trying to achieve those standards of excellence which are appropriate to, and partially definitive of, that form of activity, with the result that human powers to achieve excellence, and human conceptions of the ends and goods involved, are systematically extended" (MacIntyre, 1981: 187).

Many of the steps in the subsequent organizing processes are in the first place identified by the other participants' mindfulness. Elite players are very explicit about "working towards a goal" and "intensity in the collective situation," and the "individual contribution to the collective process."

These are frames that mean something to them in organizing the team and in team sports. Also they are familiar with relational frames like "other's knowledge about me" and "the knowledge the others hold about me without my knowing, but I want to know" (Riiser, 2003: 17). In sports, the friendly others are often the coach, who is also the norm-setter or judge of actual goals, or experienced and competent college-players, who often carry the elite player culture to the player's subjective level and consequently are important participants in intersubjective actions that transform the entire team. Maybe most important "knowers about me" are the teammates who in playing together see a player's resources, tries to understand intentions, respond and act in a way that will give more opportunities for the next pass.

Hostile others are players on the other team, understanding a player's actions, better or worse than the player himself. Their coach is always concerned with the interpretation of players' personal competences, with recognizing players' habits and the generic interdependent actions that stabilize the team.

Success is often formulated as an individual achievement, not as a question of the ability to cooperate and to contribute through the collective. A fundamental skill in all organized activity is to produce collective action, but in many places they use a rather narrow vocabulary about it. "One has to be a socialized academic to practice abstract and general reflection about relations" said an elite female handball player.[4] Everyday language is what is available for most people about the collective play, and there are few words in use about matters happening as a result of collective discussion. How, then, can a leading team produce progression and innovation?

Another female player on a successful team gives an answer.[5] She explains how changes are organized. She says that her coach may start to mention a dream she had last night. The coach tries to explain what happened in her dream team. Then she tries to instruct us about what it means in the present team, and how it could be practiced. She has a collective knowledge together with the present team. We (the players) know how the coach frames the game, and she knows how to reformulate potential situations. It happens in a very natural language which means that complex ideas are formulated in a simpler way. Whilst the team does not fully understand the message in the first place, they try the ideas out in practice for an hour or two. Afterwards they "evaluate." Here the team members remember and formulate (select) what happened and what the new things were about. In this discussion, they find collective signifiers for what they think was different and how it influenced what they assumed as agreed upon before this experience. They formulate the game in verbal terms, change theories about the game, and add some new theories.

The process is definitely innovative, but still it is kept within a set of signifiers and theories that are related to quite specific and contextual situations. The signifiers may to a great extent have the same meaning for players in this team, though it would not be quite understandable for other players, and therefore this team develops both a particular practice and a specialized meaningful language about the new type of play. The potential of the new understanding will be developed by the users as they relate to the new idea, and after some time it becomes a significant and integrated element in the teams' way to see relations during the game. A very strong innovation will change not only this team's play, but also the way other teams practice the game.

CONCLUSION

The chapter has provided a theoretical description of change processes at the micro-level. At this level, innovation and change can mean achievement of internal goods in a given practice. Achievements of internal goods can be important drivers of innovation.

The process of innovation at this level can be understood as an integral part of organizing and managing activities. Change provokes reactions and collective discussions including a search for better accounts and descriptions of the situation in question. This is not something that can be controlled at the strategic level, through formal organizations or social roles. It is an interactive process where experiences suggest that the usual practice is less qualified than it could be, and therefore the usual theory should be exchanged for better descriptions.

A better theory about what happens in a situation implies a change in understanding of the work process. Innovation becomes a series of actions leading in this direction, which means that the innovation process is an unfolding one, sometimes not predictable at all. Each innovative step is actualized through intersubjectivity, which means that each step implies changes in relations and collective meanings. Innovation implies change in worldview, which implies changes in understanding and framing of action thereafter.

Mindfulness is an integrated part of the process that leads to innovations. Mindfulness consists of scrutiny of existing expectations, differentiation of expectations when new experiences suggest them, the capability to invent new expectations that make sense of unprecedented events, frequent new appreciation of context and ways to deal with it, and identification of new dimensions of context that improve foresight. To notice unprecedented events is the trigger of the single mindful situation, and an innovative environment is dependent on such frequent notice of uncertainty, ambiguity and other signs of mismatch with former understandings of the environment.

One significant change in a relation can lead to a test of the instrumental value of the new meaning. Each of these tests may be noticed or not, and each represents a possibility for mindful organizing and intersubjectivity. In team handball, mindfulness is a necessary facilitator for the ongoing process of adjustment and search for success. Mindfulness is not the answer to difficulties, but the name of practicing virtues and collective human innovation.

NOTES

1. The term action net is a perspective on enactment and knowing, stemming from instrumentalism and pragmatism, later applied to interactionism and employed by Silverman (1970). Czarniawska (1992) reformulated the concept of action net inspired by Bruno Latour and actor network theory.
2. For a complete introduction see special issue of *Journal of Contingency and Crisis Management*, 2 (4), Basil Blackwell.
3. The actual days were 21 and 30 November and 6, 15 and 18 December.
4. Interview SS, 18 May, 2006.
5. Interview HS, 29 May, 2006.

REFERENCES

Benz, V.M. and J. Shapiro (1998), *Mindful Inquiry in Social Research*, London: Sage.
Blumer, H. (1969), *Symbolic Interactionism – Perspective and Method*, Berkeley, CA: University of California Press.
Collins, R. (1981), "Micro-translation as a theory-building strategy," in A. Cicourell and K. Knorr-Cetina (eds), *Advances in Social Theory and Methodology*, Boston, MA: Routledge & Kegan Paul, pp. 81–107.
Collins, R. (1998), *The Sociology of Philosophies – A Global Theory of Intellectual Change*, Cambridge, MA: The Belknap Press of Harvard University Press.
Czarniawska-Joerges, B. (1992), *Exploring Complex Organizations*, Newbury Park, CA: Sage.
Czarniawska, B. and T. Hernes (eds) (2005), *Actor-Network Theory and Organizing*, Malmö: Liber.
Dewey, John (1969–91), *Works*, collected by Jo Ann Boydston, Carbondale and Edwardswille, IL: Southern Illinois University Press.

Langer, E. (1989), "Minding matters: the consequenses of mindlessness-mindfulness," in L. Berkowitz, *Advances in Experimental Social Psychology 22*, San Diego, CA: Academic Press, pp. 137–73.

Mulgan, G. and D. Albury (2003), *Innovation in the Public Sector*, London: Strategy Unit, Cabinet Office.

MacIntyre, A. (1981), *After Virtue. A Study in Moral Theory*, Notre Dame, IN: University of Notre Dame Press.

Perrow, Charles (1984), *Normal Accidents: Living with High-risk Technologies*, New York: Basic Books.

Perrow, Charles (1994), "The limits of safety: the enhancement of a theory of accidents," *Journal of Crisis and Contingency Management*, **2** (4), 212–20.

Riiser, L. (2003), *Thors Afdeling – om at komme igen, om at komme videre*, Århus: Thors Forlag.

Sagan, S.D. (1994), "Toward a political theory of organizational reliability," *Journal of Crisis and Contingency Management*, **2** (4), 228–40.

Silverman, D. (1970), *The Theory of Organizations – A Sociological Framework*, London: Heinemann.

Turner, B.A. (1978), *Man Made Disaster*, London: Wykeham Publications.

Turner, B.A. (1992), "The sociology of safety," in D. Blockley, *Engineering Safety*, London: McGraw-Hill.

Weick, K.E. (1979), *The Social Psychology of Organizing*, Reading, PA: Addison-Wesley.

Weick, K.E. (1987), "Organizational culture as a source of high reliability," *California Management Review*, **24** (2), 112–27.

Weick, K.E. (1995), *Sensemaking in Organizations*, Thousand Oaks, CA: Sage.

Weick, K.E. and K.M. Sutcliffe (2001), *Managing the Unexpected – Assuring High Performance in an Age of Complexity*, San Francisco, CA: Jossey Bass.

Weick, K.E. and K.M. Sutcliffe (2006), "Mindfulness and the quality of organizational attention," in *Organization Science*, **17** (4), 514–24.

Index

Abecassis-Moedas, C. 198
absorptive capacities 176, 186, 187, 188
action 277, 299–301
    see also interaction; symbolic
        interaction
actor network theory 61–3, 255–6,
    267–8, 273
agent type 222–3
agglomeration 197
    see also clusters; disagglomeration
agora 12, 95
Albury, D. 6, 296
Amabile, Teresa M. 6, 7
Arla 112, 125, 127–8
artefacts *see* boundary objects; things
Asheim, B. 194, 195–6
association, and translation 63, 64
Åstebro, T.B. 213

becoming-realist ontology 59–60,
    80–81
behaviour 27, 28, 30, 151
Belussi, F. 198, 199
bibliotek.dk 102–5, 106
biotechnological sector 171, 174, 186,
    201, 207
    see also Medicon Valley; Medicon
        Valley Academy
Borum, F. 60, 61
bottom-up processes 28–9, 61, 75, 79,
    105–6, 107–8
boundary objects 267–8
branding 205–6
bureaucracy 142, 156
business innovation, and social
    innovation interdependence 7–11,
    12

Caby-Guillet, L. 198
Callon, Michel 61, 63, 81
capabilities 159, 249–50, 297, 298, 300,
    303, 305, 308–9

care 25, 28, 49–50, 52–3, 54, 55, 87,
    251, 299
    see also customer care; customer
        relationship management
        (CRM)
Carlile, P. 268
causality 268–9
change
    and innovation 295, 297, 308
    versus newness in innovation 276
    and organizing 296
    and service innovation 38
    and stability in Danmark Protein
        development case study 123,
        126–7
    and stability in modes of innovation
        116–17
    see also planned rational change;
        strategic change
Chesbrough, H.W. 5, 12, 13, 93–4, 95,
    113, 115, 177, 260, 292
clinical pathways 57
    see also translation metaphor in
        clinical pathways innovation
        with care
clusters 194–9, 201, 206–7
    see also Medicon Valley; Øresund
        Science Region and public
        science and industry interaction
coalition building 276
Cohen, W.M. 173, 174, 175, 176
collaboration 134
    see also coalition building;
        Management Greenhouse;
        Øresund Science Region and
        public science and industry
        interaction; partnerships;
        scientific communities;
        technological communities
collective identity 240–41, 251
collective knowledge 296, 307
collective mindfulness 300, 306–8

collective sensemaking 238, 240–42, 251, 296, 301, 302, 303, 305
collectivity 4, 5, 13, 14, 15, 16
commercialization
    and 'Mad Max Puzzle' 213, 214, 215, 216, 217, 218, 220
    and public science and industry interaction 174, 178, 184
commitment 222–3
communication 266, 270, 272
    see also conferences; conversation; cooperation; informal contacts; information; interaction; knowledge; meetings; mutual stories; publication; relationships; seminars; workshops
competencies 157, 162–3, 186
competition 36, 98, 109, 116, 140, 161
conferences 176, 183, 189
consumer choice, and convergent selection 140
consumer involvement 13, 14, 89
context 300, 305–6, 308
convergent selection 139, 140
conversation 271, 272, 273
cooperation
    in Danmark Protein development case study 120, 121–2, 125, 127, 128
    and 'Mad Max Puzzle' 214, 216, 218–19, 220
    in public library innovation 100–101, 102–6, 108, 109
    and strategies 116
    see also coalition building; collaboration; communication; Management Greenhouse; Øresund Science Region and public science and industry interaction; partnerships; public science and industry interaction; R&D cooperation; scientific communities; technological communities
coordination 287, 289–90, 291, 292
Copenhagen Hospital Corporation *see* translation metaphor in clinical pathways innovation with care
copying, in public innovation 134, 138

creative destruction 113, 239, 277
creativity 6–7, 132, 138–9, 147–8, 286–7, 290, 291, 292
critical capabilities and resources 249–50
Csikszentmihalyi, M. 6–7, 88, 131
cues 240, 241, 242, 243, 248–9, 250, 251
customer care 43, 49–51, 53–4, 55
customer involvement
    in customer relationship management implementation 49, 50–51, 54
    and industrial R&D 171
    in service innovation 12, 29–30, 33–4, 35, 36, 39, 43
    and technology development 173
customer relationship management (CRM) 48–55
customers
    and customer relationship management implementation 51, 52, 53, 54
    in e-realtor sense caring case study 246, 247, 251
    and market knowledge 50, 276
    in Sidecar project and durability of innovation case study 261

Dahlin, K.B. 213
danbibbase 101–3
Danish Bibliographic Centre (DBC) 97, 101–2, 103, 104, 105, 106, 108
Danish National Library Authority 97, 98, 104, 105, 106, 108
Danish Union of Librarians 100
Danmark Protein development case study
    conclusions 127–8
    entrepreneurship 118–20, 126
    exploitation 121–5, 126–7, 128
    exploration 118–23, 126, 127, 128
    and globalization 124–5, 127–8
    open innovation 124–5, 127
    strategic reflexivity 127, 128
    techno-economic innovation 121–4, 126–7
    whey proteins as a business 113–15, 118–19
de-institutionalization 64

decentralization, of municipalities
    143–4, 146–8, 152
decision-making 29, 32, 35, 36, 37, 40,
    41, 136–7
demand pull model 170
development 29, 37–8, 41, 173
Dewey, John 298–9, 301
diamond model of innovation 200
diffusion
    in actor network theory 62
    and entrepreneurship 92
    and importance 4
    and innovation with care 8, 13
    and innovation with care in public
        institutions 89, 90
    and open innovation 113
    and organizational innovation 276
    and public library innovation 97, 106
Diginet Øresund 180, 185
DiMaggio, P.J. 87, 88, 90, 92, 145
disagglomeration 197
divergent selection 139–40
diversity 3–4, 5, 13, 14, 15, 148, 159
division of labour 65, 66, 67, 69–70,
    73, 74, 78, 79
domain of innovation 88, 97–8, 101–6,
    107, 109, 131, 137, 138–9
Drejer, I. 276
durability 63, 255, 272, 273
    see also Sidecar project and
        durability of innovation case
        study

e-realtor sense caring case study
    background 243–4
    collective sensemaking 251
    cues 241, 243, 248–9, 250, 251
    market disequilibrium and idea
        conception 245, 248, 250
    organizational sensemaking and new
        networks 248, 250
    simplicity and the Internet 245–6
    strategizing 248–9
    trust relationships 247, 251
economic criteria, and public
    innovation 135, 136
economic efficiency, and reform of
    municipalities 144
educational organizations 145, 148, 149
    see also universities

employee involvement
    and customer relationship
        management implementation
        49, 52–3
    in innovation 14
    in public innovation 152, 157, 158,
        164
    in public innovation with care 89,
        133, 134, 135, 136, 138–9
    in public library innovation 105–6,
        107–8, 109
    in service innovation 12, 28, 29, 32,
        33, 34–5, 36, 37, 38–9, 40–42, 43
employees *see* employee involvement;
    frontline employees; librarians;
    manager involvement
empowerment 42, 282–3, 284, 287, 288,
    289–90, 291, 292
    see also power
enactment 241, 249, 250
entrepreneurship
    in Danmark Protein development
        case study 118–20, 126
    and globalization 116
    paradigm 115
    in public innovation 134, 152–3
    in public library innovation 105, 106,
        107–8
    roles 11, 276, 291
    in service innovation 30, 34, 36
    in Sidecar project and durability of
        innovation case study 257
    theories 6, 11, 91–2, 112, 239, 277
    see also intrapreneurship; Kearns,
        Robert and the intermittent
        windscreen wiper case study;
        'Mad Max Puzzle'; individual
        perspective
environmental factors
    in Danmark Protein development
        case study 120, 123, 126, 128
    and innovation 113, 114–15
    and open innovation 115–16
    and public innovation 155
    and strategic reflexivity 116, 117
    and technological innovation 199–200
episodes 63–4
    see also translation metaphor in
        clinical pathways innovation
        with care

equivoque technologies 279
Etzkowitz, H. 177–8, 200, 203
expectations, and mindfulness 300,
   304–5, 308–9
experience
   in handball match and mindful
      innovation case study 297, 301,
      302, 303, 304–5
   and innovation 295, 296, 298, 308–9
   new technology 278, 279, 291, 293
   and sensemaking 4
explicit knowledge 38, 206, 261
exploitation 6, 8, 12, 117, 121–5,
   126–7, 128
   see also 'Mad Max Puzzle'
exploration 6, 8, 13, 118–23, 126, 128
external sources 93–4, 105, 106, 115,
   116, 260, 261–2, 268–71

fashion designers *see* Sidecar project
   and durability of innovation case
   study
field of innovation 98–9, 101–6, 107,
   108, 131, 133, 136, 137, 138
fluidity 255, 264
Ford Motor Company 226, 227–30
frames 307, 308
   see also technology frames
Frederiksberg Hospital *see* translation
   metaphor in clinical pathways
   innovation with care
frontline employees 12, 13, 30–31, 39,
   52–3, 54
Fuglsang, Lars 5, 32, 90, 95, 113, 114,
   116, 267, 268, 269
functionality, technology 278–9

gains versus losses 224–5
Gash, D.C. 278
Giddens, A. 63, 277
globalization 116, 124–5, 127–8, 143,
   194, 206–7
Google 104–5, 108, 140
Gottardi, G. 197–8
government 177, 183, 187–8, 189,
   199–200, 203–5, 221

handball match and mindful
   innovation case study
   coach 301–4

collective mindfulness 306–8
mindfulness characteristics 304–6
virtue and selection 297–8
Hauknes, J. 57, 150, 152–4
health care, and New Public
   Management 57–9
health care innovation 58–9, 76
   see also translation metaphor in
      clinical pathways innovation
      with care
high tech industries 171
   see also biotechnological sector;
      ICT sector; pharmaceutical
      sector
high tech platforms, in Øresund
   Science Region 183, 184–5, 186
homogenization 87, 89, 90, 92–3

ICT (information and communication
   technologies) 27, 49, 194, 197–9,
   201, 205–7
   see also e-realtor sense caring case
      study; Internet; Øresund IT
      Academy; ProjectWeb
      sensemaking case study
ICT sector 186
ideas
   and causality 268–9
   in e-realtor sense caring case study
      244, 245
   in service innovation 28–9, 32, 33,
      34–6, 37, 40–41
   in Sidecar project and durability of
      innovation case study 256–7,
      268–9
identity 240–41, 248, 251, 272
implementation
   as barrier to public innovation 157
   customer relationship management
      49, 50–53
   defined 58
   importance in organizational
      innovation 275
   metaphor 60–61, 75, 76
   new technology 278–80
      (*see also* ProjectWeb sensemaking
         case study)
   service innovation 29, 38–40, 41, 43
   theories 58
importance, concept of 4, 14–15

in-side out scripts 113, 114–15, 116, 123
individual perspective 176, 213, 290–91, 292, 293, 296
    see also entrepreneurship; Kearns, Robert and the intermittent windscreen wiper case study; 'Mad Max Puzzle'
industrial localization 193–4, 195, 196, 207
    see also clusters; Medicon Valley; Øresund Science Region and public science and industry interaction
industry
    and innovation 199–200
    and R&D cooperation 171
    triple helix model of innovation 177–8, 183, 187–8, 189, 200, 203–5
    see also biotechnological sector; high tech industries; ICT sector; large firms; low tech industries; medium tech industries; Øresund Science Region and public science and industry interaction; pharmaceutical sector; public science and industry interaction; small firms
informal contacts 176, 183, 186–7
information 239, 240, 245
information ecologies, and new technology appropriation 276–7, 279
innovation
    concept 3, 88, 295–6
    and creativity 6–7
    definitions 6, 9, 49, 57, 150–51
    diamond model 200
    theories 11–13, 112–13
    triple helix model 177–8, 183, 187–8, 189, 200, 203–5
    types 26, 57, 151
innovation culture 158, 161–2
innovation networks 30–31, 36
innovation partnerships 106
innovation with care 4–11, 88, 89–90, 108, 254
inquiry 298–9

institutional agency, and public innovation 145–6
institutional development, municipalities 146–8
institutional innovation 92–3, 100–101, 107, 109
institutional theory 90, 92–3, 144–6, 159
institutionalization 64, 142, 147, 152
'institutions for professionals' 139
    see also universities
interaction 3, 5, 6, 132
    see also coalition building; collaboration; communication; cooperation; partnerships; public science and industry interaction; relationships; scientific communities; symbolic interaction; technological communities
interest, in ProjectWeb sensemaking case study 285, 286, 287–8, 289, 290
intermittent windshield wiper case study 225–31
internal sources, in open innovation 93–4, 116, 269
international pool of knowledge 183
international scientific communities 176, 203, 204–5, 207
Internet
    and clusters 197, 198
    and innovation with care 13
    and Medicon Valley 199, 205–7
    and public innovation with care 139, 140
    and public library innovation 97, 98, 99, 102–6, 107, 108
    see also e-realtor sense caring case study
intersubjectivity 295–6, 300, 308, 309
intrapreneurship
    and networks 276
    in ProjectWeb sensemaking case study 289–90, 291–2
    in service innovation 31–2, 33, 34, 35, 37, 40–41, 42
invention service firms 221
inventors *see* entrepreneurship; intrapreneurship

involvement 3–4, 14, 29–30, 266
    see also consumer involvement;
        customer involvement;
        customer relationship
        management (CRM); employee
        involvement; manager
        involvement; strategic reflexivity
        model of service innovation;
        supplier involvement;
        translation metaphor in clinical
        pathways innovation with care;
        user involvement
isomorphism, and institutional
    innovation 92, 93
Italy 195, 197, 198, 201

Japan 119–20, 122–3, 126

Kahneman, D. 222, 224
Kali Chemie AG 120–22, 124, 126
Kanter, R.M. 276, 291
Kearns, Robert and the intermittent
    windscreen wiper case study
    225–31
Kirzner, Israel M. 91, 239, 277, 280
Klausen, K.K. 146, 148–50, 159
know-how *see* 'Mad Max Puzzle'
knowledge
    commercialization 174, 178, 184
    and mindfulness 299–301
    in ProjectWeb sensemaking case
        study 285, 286, 287, 289, 290,
        291
    and trust relationships 247
    see also collective knowledge;
        explicit knowledge;
        information; knowledge
        sharing; knowledge transfer;
        learning; local knowledge; 'Mad
        Max Puzzle'; market
        knowledge; new knowledge;
        pool of knowledge; tacit
        knowledge; understanding
knowledge sharing 151, 161
knowledge transfer 177–8, 181–2, 201,
    205–7
    see also R&D cooperation
Koch, P. 57, 150, 152–4
Komnunale Tjenestemænds
    Organisation (KTO) 163–4

Krabbe, Jørgen 118–19, 120, 121, 122,
    126
Kumar, K. 197, 199, 206

large firms 184–5
    see also 'Mad Max Puzzle'
Latour, B. 59, 61, 255–6, 264, 269
Laursen, K. 173, 176–7
Law, John 255
Leamer, E.E. 197
learning 80, 278, 279
    see also information; knowledge
        sharing; knowledge transfer;
        new knowledge; organizational
        learning; pool of knowledge;
        understanding
Leydesdorff, L. 177, 200, 203
librarians 98, 102, 105–6, 107–8, 109
Library Association 100
library cause 100, 101, 109
library education 97, 101
Library Watch 105–6, 107, 108
licensing, and 'Mad Max Puzzle' 214,
    220, 221, 230
local authorities 188, 189
local communities, and public
    innovation with care 89–90
Local Government Denmark (LGDK)
    163–4
local knowledge 69, 74, 77
local learning 80
local managers 157, 163–4
local processes, in clinical pathways
    innovation with care 76, 77, 80
lone inventors *see* entrepreneurship;
    Kearns, Robert and the
    intermittent windscreen wiper
    case study; 'Mad Max Puzzle';
    individual perspective
losses versus gains 224–5
low tech industries 171, 176–7
low tech platforms, in Øresund Science
    Region 183–4, 185–6, 187

MacIntyre, Alistaire 297–8, 306
'Mad Max Puzzle'
    described 214–17
    Kearns, Robert and the intermittent
        windshield wiper case study
        225–31

and positioning and the negotiation
    agenda 223–5
and positioning over time 217–20
and positioning through negotiation
    skills 222–3
and positioning with a multitude of
    agents 220–22
Maignan, C. 197
management 157–8, 160–63
    see also customer relationship
        management (CRM); local
        managers; manager
        involvement; middle managers;
        New Public Management; top
        management
Management Greenhouse 146, 148,
    153, 163–4
manager involvement
    and customer relationship
        management 51, 52–3, 54
    in public innovation 157, 158, 163–4
    in public innovation with care 89,
        138–9
    in service innovation 28, 29, 30, 32,
        33, 34, 35, 36, 39, 40–41
manufacturing innovation 26, 27, 28,
    30–31, 36
March, J.G. 117
market disequilibrium 91, 239, 240,
    245, 248, 250, 251
market knowledge 50, 276
market pull model 115
marketing 32, 33, 37, 52
markets
    and convergent selection 140
    Danmark Protein development case
        study 119–20, 122–3, 124–5, 126
    and innovation 8, 9, 115
    and R&D cooperation 171
    and service innovation acceptance
        38, 39
Marshall, A. 195, 196
materiality 271, 272
materialization 256–7, 259–60, 262–4,
    265, 271
    see also boundary objects;
        materiality; things
MD-Foods 118–19, 120, 122, 126
media attention, and divergent
    selection 140

Medica case study *see* ProjectWeb
    sensemaking case study
Medicon Valley 199, 201–2, 205–7
Medicon Valley Academy 180, 182,
    183, 184, 187, 201, 203–7
medium tech industries 171
meetings 176, 183, 187, 206
meta-innovation 146, 163–4
metaphors 12, 59, 60–61, 75–6, 95
middle managers 28, 39, 163–4
mindful cooperation 116
mindful innovation 295
    see also handball match and mindful
        innovation case study
mindfulness 299–301, 308–9
modes of innovation 90–96, 115–17
Møller, Jørn Kjølseth 148, 153, 162
Morgan, Gareth 60, 61
motivation 9–10, 38, 155
Mulgan, G. 6, 12, 296
municipalities 143–4, 146–8, 152, 153,
    163–4
mutual stories 266
mutual understanding 271, 272, 273

Nano Øresund 180, 183, 184–5
negotiation
    and innovation 271, 273–4
    and 'Mad Max Puzzle' 214, 215–16,
        217–18, 220, 221, 222–5
    see also Kearns, Robert and the
        intermittent windscreen wiper
        case study
neo-instititutionalism 92–3, 144–6,
    159
networks
    and critical capabilities and
        resources 249–50
    in Danmark Protein development
        case study 125
    in e-realtor sense caring case study
        246, 248–9, 250, 251
    importance in innovation 237–40,
        249–50
    and intrapreneurs 276
    in public innovation 153–4,
        160–63
    and public science and industry
        interaction 177
    and sensemaking 240–41, 242

in Sidecar project and durability of
innovation case study 260–61,
265–6
see also Management Greenhouse;
Øresund Science Region and
public science and industry
interaction
new customers 51, 52
new knowledge 174–6, 182–3, 186
New Public Management 57–9, 64, 76,
77–8, 79, 80, 147, 148
new technology 276–7, 278–80, 292
see also biotechnological sector; ICT
(information and
communication technologies);
ICT sector; Internet;
ProjectWeb sensemaking case
study
newness 6, 276
norms 116, 117, 138, 171
Nowotny, Helga 12, 95

objects *see* boundary objects;
materiality; materialization;
things
open innovation
defined 93–4
described 115–16
and diffusion 113
external sources 93–4, 105, 106, 115,
116, 260, 261–2, 268–9
and innovation with care 94–5
internal sources 93–4, 116, 269
and new technology 292
and public library innovation 88,
105, 106, 108–9, 110
in Sidecar project and durability of
innovation case study 260–62
Øresund Environment Academy
(ØEA) 180, 184, 185
Øresund Food Network (ØFN) 180,
184, 185–6, 187
Øresund IT Academy 180, 183, 184,
187
Øresund Logistics (ØL) 180, 184, 185
Øresund Science Region and public
science and industry interaction
background to Øresund Science
Region 179–80, 202–3
channels of interaction 183–6, 189

new knowledge versus pool of
knowledge 182–3
new structures between public
science and industry 186–7
public science and industry relations
as interactive process 181–2
strengths and weaknesses 189
and triple helix model 183, 187–8,
189
see also Medicon Valley; Medicon
Valley Academy
organizational culture 158, 159,
160–63, 248, 251
organizational entrepreneurship *see*
intrapreneurship
organizational innovation 275–6
see also ProjectWeb sensemaking
case study
organizational learning
in Danmark Protein development
case study 114, 121–2, 123,
126–7
in public innovation 151, 161–3
in service innovation 30, 38, 41–2
see also Management Greenhouse
organizational sensemaking 248,
278–80
organizing 295, 296, 300–301, 309
Orlikowski, W. 278, 279
out-side in scripts 113, 114, 116
outsourcing 115, 260

Parsons, W. 60–61
partnerships 106, 118–19, 120–22, 139
patenting 213, 214, 217, 218, 220, 221,
226, 227, 228, 229, 230
path dependency, in public innovation
142, 144–5, 148, 159
Pedersen, John Storm 132, 144
Perrow, Charles 299
pharmaceutical sector 171, 174, 186
see also ProjectWeb sensemaking
case study
planned rational change 58–9, 61, 75–6
pokerness, in open innovation 94–5
political-administrative systems, and
public innovation 134, 136–7, 137,
138
pool of knowledge 174–6, 182–3
Popper, Karl R. 9, 89, 132

Porter, M.E. 193, 194, 195, 196, 200
positioning 4, 15, 126, 265–6
    see also 'Mad Max Puzzle'
Powell, W.W. 87, 88, 90, 92, 145
power 270, 272, 276
    see also empowerment
practice 296, 297–8
privatization 147–8
process innovations 26, 151
product innovations 26, 151, 280, 292
professional criteria, and public
    innovation with care 135, 136,
    138–9
professionals 134, 136–7, 138, 156,
    260–61
profitability 9, 53–4
project assistants 282, 283, 284, 285,
    286, 287–8, 289–90, 291–2
project groups, in service innovation
    29, 37–8, 41
ProjectWeb sensemaking case study
    discussion 290–92
    empirical setting 280–84
    implementation of ProjectWeb
    284–5
    intrapreneurship in implementation
        of ProjectWeb 289–90, 291–2
    Project 1 285–6, 289, 290
    Project 2 286–7, 289, 290, 291
    Project 3 287–9, 290, 291
promotion of ideas, in service
    innovation 34
prototypes 220, 221, 226, 228, 259,
    260, 262–4
public innovation
    diversity and capability 148, 159
    domain of innovation 137, 138
    drivers and barriers 8, 154–7
    field of innovation 133, 136, 137,
        138
    versus innovation with care in survey
        analysis 132–3, 134–9
    and institutional agency 145–6
    Management Greenhouse 146, 148,
        153, 163–4
    management of 157–8
    meta-innovation 146, 163–4
    modes 90–96
    municipalities, reform 143–4, 146–8
    networks 153–4, 160–63

and organizational culture 158, 159,
    160–63
and organizational learning 151,
    161–3
path dependency 142, 144–5, 148, 159
strategies 151–4
survey 133–4
types 57, 151
public innovation with care 89–96,
    132–3, 134–9
public libraries
    competition 98, 109
    purpose 98–9
    social and strategic arenas 97–9,
        101–6, 107, 108
    as social institutions 99–101
public library innovation
    entrepreneurship 105, 106, 107–8
    institutional innovation 100–101,
        107, 109
    open innovation 88, 105, 106, 108–9,
        110
    opportunities 101–6
    and selection 97, 104, 105, 106, 107,
        108, 109, 110
    strategic reflexivity 102, 104, 108–9,
        110
    and variation 97, 105, 106, 107, 108,
        109, 110
public science 171
public science and industry interaction
    as 'mechanical' 169
    public science as new knowledge or
        pool of knowledge 174–6
    public science contribution to
        technology development 173
    R&D cooperation 171–2
    structural factors 176–7
    transfer contrainteraction 173-4
    triple helix model 177–8, 183, 187–8,
        189, 200
    see also Øresund Science Region and
        public science and industry
        interaction
publication 171, 176

qualifications 279–80, 289, 290
qualitative approaches to quality
    development 65, 68, 70, 71, 72,
    73–4, 77, 79

quality
  in Danmark Protein development
    case study 121, 122, 123
  and reform of municipalities 144
  in service innovation 26, 27–8, 41–2
  in translation metaphor in clinical
    pathways innovation with care
    65, 68, 70–72, 73–4, 77, 79
quantitative approach to quality
  development 65, 68, 70–72, 73, 74,
  77, 79

R&D cooperation 171–2
recognition, and innovation with care
  131, 139
relationships 267
reproduction 35, 276
retrospection 241, 249, 250, 296
RobinHus *see* e-realtor sense caring
  case study
roles
  and customer relationship
    management implementation 52
  in entrepreneurship 11, 276, 291
  in project groups 38
  in public innovation 158
  in public library innovation 97, 98,
    105–6, 109
  in service innovation 35, 38, 40–41,
    42
Rosenfeld, S.A. 196, 201
routines 11, 12, 116, 117, 156

safety 299
Salter, A. 173, 176–7
scaling up 8, 13, 15, 106, 132
Schelling, T. 222, 223, 225
Scheuer, John Damm 58, 60, 61, 62, 63
Schumpeter, Joseph A. 6, 7, 9–10, 11,
  91, 92, 112, 113, 239, 276, 277,
  291
Schumpeter I, II and III 11–13, 94
science, public *see* public science;
  public science and industry
  interaction
science push model 170
scientific communities 171, 176, 203,
  204–5, 207
Scott, W.R. 92, 145
Scupola, Ada 194, 198–9, 201, 205

selection
  and collective mindfulness 300
  convergent 139, 140
  divergent 139–40
  and entrepreneurship 91–2
  in handball match and mindful
    innovation case study 298,
    307–8
  and importance 4
  and innovation 6, 8, 13, 14, 15, 16
  and innovation with care 8, 131
  and innovation with care in public
    institutions 89, 90, 131
  and open innovation 94–5, 105
  and public innovation with care 132,
    135, 136, 137, 138, 139
  and public library innovation 97,
    104, 105, 106, 107, 108, 109,
    110
  in Sidecar project and durability of
    innovation case study 260, 261
  and strategic reflexivity 95
seminars 183, 187, 189
sensemaking
  concept 4, 15–16
  properties 240–42, 278
  and strategizing 250, 251
  and strategy 250, 251
  and technology 278–80
  see also collective sensemaking; e-
    realtor sense caring case study;
    organizational sensemaking;
    ProjectWeb sensemaking case
    study; translation
serendipity 240, 245
service concepts 32–3, 34
service innovation
  and customers 12, 29–30, 33–4, 35,
    36, 39, 43
  empirical examples 40–43
  and employees 12, 28, 29, 30-31, 32,
    33, 34–5, 36, 37, 39, 40–42, 43
  nature of 26–8
  and open innovation 115–16
  organization 28–31
  see also strategic reflexivity model of
    service innovation
services, defined 25, 26
Sidecar project and durability of
  innovation case study

aligning space 264–5
aspiring fashion designers 257–9
entrepreneurship 257
innovation as persistent action 268–9
involvement 266
materialization of ideas 256–7, 265
materializing the of idea of a sweater 259–60, 262–4
and mutual understanding 271
open innovation 260–62
and positioning 265–6
selection 260, 261
strategic reflexivity 268, 269–71
turning innovation into premises for people's own actions 267–8
small firms 185, 187
social and strategic arenas 90, 95, 96, 101–6, 107, 108, 139
social engineering 96
social innovation, and business innovation interdependence 7–11, 12
social institutions 99–101
social mechanisms of innovation 8, 9
social skill, in actor network theory 255
society, and public innovation with care 89
software 49
see also ProjectWeb sensemaking case study
stability 64, 116–17, 123, 126–7, 255, 264–5, 273, 296
Steinfield, Charles 194, 198–9, 201, 205
Storper, M. 197
strategic arenas 148–50
see also social and strategic arenas
strategic change 122–4
strategic decision-making 136–7
strategic planning 238, 239, 240–41
strategic positioning 126
strategic reflexivity
in Danmark Protein development case study 127, 128
external sources 268–71
and innovation with care 95–6
involvement and care model of service innovation 31–4

public library innovation 102, 104, 108–9, 110
in Sidecar project and durability of innovation case study 268, 269–71
and stability and change balance 116–17
strategic reflexivity model of service innovation
care concept 31
development phase 37–8, 41
idea phase 34–6, 40–41
implementation phase 38–40, 41, 43
strategic reflexivity concept 31–4
strategizing 238, 248–9, 250, 251
see also handball match and mindful innovation case study
strategy 116, 151–4, 250, 251, 278–9
success 6, 307
Suchman, Lucy 258
Sundbo, J. 5, 6, 27, 28, 31, 32, 35, 36, 38, 40, 41, 90, 95, 113, 114, 115, 116, 239, 267, 269, 275, 277, 284, 291
supplier involvement 36, 173
Sutcliffe, K.M. 300
Sweden *see* Øresund Science Region and public science and industry interaction
symbolic interaction 241, 295–6
see also handball match and mindful innovation case study

tacit knowledge 38, 42, 206
Tann, David 226, 227, 228, 229
Taylor, Frederick Winslow 58–9
techno-economic innovation 115, 121–4, 126–7
technological communities 176
technological innovation 27, 199–200
technology, sensemaking 278–80
technology development, public science contribution 173
technology frames 241, 278–80, 284–5, 286, 287, 289, 290–91
technology in use 278–9, 292
see also ProjectWeb sensemaking case study
technology mediators 279–80, 291
see also project assistants

technology push model 115
things 62, 63, 64, 255–6, 272
    see also boundary objects;
        materiality; materialization;
        Sidecar project and durability
        of innovation case study
Thompson, J.B. 117
top-down processes 28–9, 35, 61, 75,
    102
top management
    and customer relationship
        management 51, 52
    in Danmark Protein development
        case study 123
    and health care innovation 28–9, 32,
        34, 35, 36, 37, 40–41, 43
    and open innovation 96
    and public innovation 157, 160, 161,
        163–4
    and service innovation 4, 28–9, 32,
        34, 35, 36, 37, 40, 43
translation
    in actor network theory 62, 273
    as association 63, 64
    concept of 59
    defined 63
    episodes 63–4
    metaphor 16, 59, 75, 76
    in Sidecar project and durability
        of innovation case study
        262–4
    see also sensemaking; translation
        metaphor in clinical pathways
        innovation with care
translation metaphor in clinical
    pathways innovation with care
    conclusions 78–80
    discussion 75–8
    episode 1 65, 66–8, 73
    episode 2 65–6, 68–70, 73–4
    episode 3 66, 70–73, 74
    introduction to case 64–5
    practical consequences 79–80
transparency, and public innovation
    with care 135

triple helix model of public science and
    industry interaction 177–8, 183,
    187–8, 189, 200, 203–5
trust relationships 95, 247, 251
Tversky, A. 222, 224

understanding 284–5, 286, 287, 289,
    290, 291, 292, 293
    see also mutual understanding
universities 171, 175, 177, 183, 187–8,
    189, 200, 203–5
    see also Øresund Science Region and
        public science and industry
        interaction
user involvement
    in ProjectWeb sensemaking case
        study 281, 282–3, 285–6, 287,
        288–9, 290, 291–2, 293
    in public innovation with care 135, 136
    in public library innovation 102–4,
        108
    in technology development 173

variation
    and entrepreneurship 92
    and innovation 6, 8, 13, 14, 15
    and innovation with care 89, 131
    and open innovation 94, 105
    and public innovation with care 132,
        133, 135, 136, 137–9
    and public library innovation 97,
        105, 106, 107, 108, 109, 110
    and strategic reflexivity 95–6
Vikkelsø, S. 59, 61, 76
Vinge, S. 59, 61, 76
virtue 297–8

Weber, Max 117, 156
Weick, Karl E. 15, 240–42, 278, 279,
    295, 298, 299, 300
whey and whey proteins 113–15,
    117–18
workshops 183, 189

Zander, Karen 57, 77

.